Industrial Locomotives of South Staffordshire

INDUSTRIAL LOCOMOTIVES of SOUTH STAFFORDSHIRE

Compiled by R M Shill

Series Editor: R K Hateley

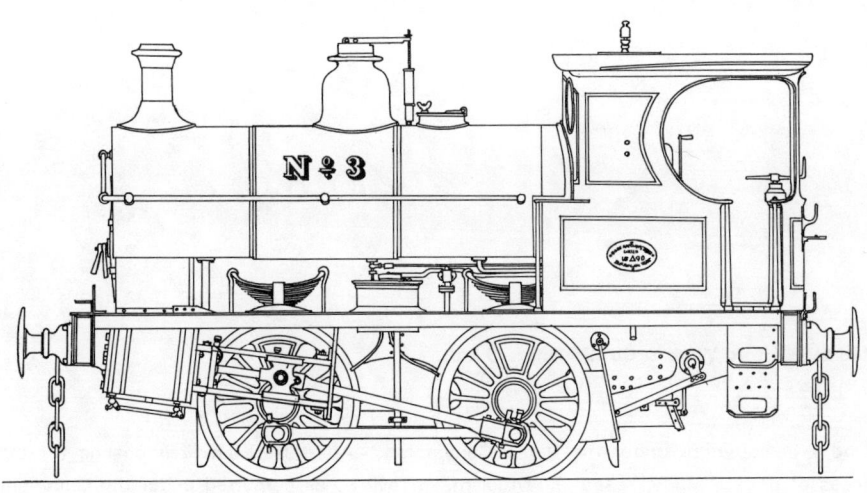

No.3 (TW 609/1890) at Bass, Ratcliff & Gretton Ltd, Burton-on-Trent.
Drawing by Chris Fisher from original survey and preliminary drawing by Roger West

INDUSTRIAL RAILWAY SOCIETY

Published by the INDUSTRIAL RAILWAY SOCIETY

at 47 Waverley Gardens, LONDON, NW10 7EE

© **INDUSTRIAL RAILWAY SOCIETY** 1993

ISBN 0 901096 77 6 (hardbound)

ISBN 0 901096 78 4 (softbound)

Printed by Clifford Ward & Co (Bridlington) Ltd.

This book is copyright under the Berne Convention. Apart from any fair dealing for the purposes of private study, research, criticism, or review, as permitted under the Copyright Act, 1911, no portion may be reproduced by any process without the written permission of the publisher.

CONTENTS

	INTRODUCTION	7
	EXPLANATORY NOTES	11
SECTION 1	Area Maps detailing locations of industrial railways within South Staffordshire.	19
SECTION 2	An Alphabetical list of all firms known to have used locomotives within South Staffordshire.	37
SECTION 3	A list of all British Coal/ National Coal Board locations within South Staffordshire.	85
SECTION 4	Details contractors lines within South Staffordshire.	119
SECTION 5	Preserved locomotives within South Staffordshire	125
SECTION 6	A summary of those locations such as engineers yards and dealers.	129
SECTION 7	Non locomotive lines within South Staffordshire.	133
SECTION 8	Index of Locomotive Builders.	146
SECTION 9	Index of Locomotives in manufacturer order.	148
SECTION 10	Index of Proprietors.	165
SECTION 11	Index of Locations.	171
SECTION 12	A selection of photographs concerning industrial locomotives within South Staffordshire .	173

INTRODUCTION

Staffordshire occupies a key position in the Midlands and encompasses the crossroads of the nation's transport network. Roads, railways and canals; all were built to serve the growing needs of industry.

The growth of coal mining and iron manufacture through the eighteenth and nineteenth centuries provided powerful reasons for better transport than then existed. Muddy tracks hampered the conveyance of goods and increased their prices in the market place. The building of canals and railways promoted industry, for they offered cheap carriage, even to previously inaccessible places.

For Staffordshire this change began during the eighteenth century when James Brindley engineered the first canals. After his success with the Bridgewater canal, Brindley embarked on grander projects. These included the Grand Trunk Canal, which linked the River Mersey with the River Trent and the Wolverhampton canal, which split from the Grand Trunk to join the River Severn. Both of these canals crossed Staffordshire and stimulated its industrial growth.

Where terrain dictated that direct canal links were not possible, tramroads or tramways were built to carry minerals to the nearest canal wharf. In time, more lengthy 'railways' were considered.

Local industry benefited tremendously with the coming of the railways. In addition to public railways many private lines were constructed to aid in the transport of minerals, industrial products and manufactured goods. Thus was born the industrial railway.

Staffordshire has had its share of industrial railways, perhaps more than many counties. In previous Pocketbooks devoted to this area the emphasis was placed on the industrial locomotives and little space was given to the railways on which they worked. This present offering seeks to redress this omission and incorporates items of historical background together with details concerning non-locomotive worked tramways and railways. Transitory railways such as contractors lines are also listed where known, while other modes of transport such as canals are mentioned when they are of direct relevance.

This volume continues the tradition of the coverage of the industrial locomotive within Great Britain, a prime aim of the Industrial Railway Society, and deals with the industrial railways and locomotives to be found within South Staffordshire. The area has been dealt with in two previous books, both published under the auspices of the Birmingham Locomotive Club. Perhaps understandably the BLC's very first "pocket book" published in 1947 was 'Industrial Locomotives of the West Midlands', a modest 42 page booklet which covered the industrial locomotives of Staffordshire, Worcestershire and Warwickshire. A 90 page revised and enlarged edition - with the same title but designated Pocket Book A - was published by the BLC's Industrial Locomotive Information Section in 1957. The BLC-ILIS subsequently changed its name to the Industrial Railway Society in 1968. Since 1957 much additional information has come to light and opportunity has been taken to incorporate this in the text. In particular a concerted effort has been made to examine archive material in order to gain a better picture of the industrial scene and also detail forgotten railways.

So much new material has been added that it is necessary to confine this volume to South Staffordshire. A complementary volume, covering the West Midlands County, was published in 1992. Other sections for those sections of Warwickshire and Worcestershire not incorporated into West Midlands and for North Staffordshire will appear under separate covers.

South Staffordshire is a mixture of quite different communities ranging from rural to industrial. Towns such as Burton-upon-Trent and Tamworth, lying on the eastern boundary of the county, are included. In fact Tamworth was once split and the boundary between Staffordshire and Warwickshire ran through the centre of the town. The boundary has since been moved eastwards and Glascote and Wilnecote are now in Staffordshire. Cannock Chase occupies the large central tract of South Staffordshire, while to the west the remainder of the county curves around the edges of the Black Country to terminate at Kinver.

Much of the present volume deals with the coal industry and in particular the Cannock Chase coal field. Though coal has been mined on Cannock Chase for centuries this was on a relatively small scale Not until the mid-nineteenth century were the deep pits developed and their rich seams of coal exploited.

The manufacture of iron was once important on Cannock Chase. Wood from the local forests was converted into charcoal to fuel iron-smelting bloomeries located on windswept parts of the Chase such as Beaudesert. These were sourced with iron ore brought by pack horse from mines in the Cheslyn Hay and Walsall areas. During the Industrial Revolution coke replaced charcoal as fuel and the iron making process moved south to the Black Country.

By 1850 the shallow coal mines at Brereton, Brownhills, Essington and Great Wyrley were exhausting their reserves. Deeper measures were sought and several were found in the Cannock area. Within fifteen years several new mines had been established around Cannock and Hednesford and these were able to supply large quantities of coal. This newly found mineral wealth attracted the railway companies of the day and both the London & North Western Railway and the Midland Railway built branches to serve the new pits. Links were also made by water. While other canals were in decline, the Birmingham Canal Navigation completed a new canal to Hednesford to serve the centre of the coal field.

Whole new communities grew up from nothing. Such was the rapid development of the Cannock Chase coal field. The decline happened in the 1950's and 1960's almost a quickly as the growth of a century earlier. Many of the mines acquired by the National Coal Board were closed within fifteen years. Now all but one have gone.

There was also a small coal field around Tamworth and Glascote, geologically part of the Warwickshire coal field, with mines that were a strange blend of the old and the new. Some mines near Wilnecote were at work at the start of the nineteenth century, while Amington was a much later sinking. The clay found in this district was almost as useful as the coal and was responsible for the establishment of several brickworks.

By contrast Burton-upon-Trent's main asset was its waters, which have favoured the brewing of beer for generations. Burton's first industrial transport link was the River Trent which allowed carriage of the town's produce down to the sea. Consequently the town's first breweries were to be found alongside the bank of that river. Canals and railways later came to serve the town, but it was not until 1861 that plans were made to connect the existing breweries with the railways. By that time the Midland Railway served Burton and within a few years a mesh of railway lines covered the town. Much of the track belonged to the Midland Railway but there was also a considerable mileage of private railway laid by the brewery companies.

Gradually much of the land between the Trent and the Midland railway was taken by different breweries, their maltings and cooperages. Railways became the main method of transport within Burton's breweries and remained so for the first half of the present century. At one time there were thirty two level crossings within Burton and fifty signal boxes existed to control the traffic.

Several engineering firms also existed within this town, largely to serve the local mining and brewing industries. One, Thornewill & Warham, constructed a variety of industrial plant which included a number of steam locomotives. Successive companies under the Baguley title specialised in producing petrol and diesel locomotives.

For present day enthusiasts it is a sad fact that much of the text of this book concerns itself with the past. Apart from preserved railways and locomotives, very little has survived through to the present day. Indeed at the time of writing, only one working line remains - that serving Littleton Colliery. This trend is one repeated throughout Britain's industrialised regions. Regardless of where one visits - Tyneside or the Clyde Basin, South Wales or Staffordshire, or even the 'modern' Yorkshire and Nottinghamshire coal field - industrial railways are rapidly in decline. Something of their memory, however, has been captured for future railway students in the IRS county series of Handbooks.

ACKNOWLEDGEMENTS

The compilation of this handbook has involved the co-operation of many people. Special thanks is given to the late W.K. (Bill) Williams, Gordon Green, P.G. Hindley and Eric Tonks whose help has been invaluable in producing the draft. I am also most grateful to the following who all have contributed information

C.A.Appleton	J.Haddock	J.A.Peden
A.C.Baker	E.W.Hannan	L.W.Perkins
V.J.Bradley	M.Hale	K.P.Plant
D.Clayton	R.K.Hateley	S.T.Pritchard
R.D.Darvill	late S.H.P.Higgins	J.T.Rhead
B.Dickens	P.G.Hindley	late R.T.Russell
C.G.Down	F.Jones	B.Rumary
P.Ellis	F.Jux	F.D.Smith
A.R.Etherington	J.B.Latham	C Shepherd
J.Evans	M.J.Lee	late B.D.Stoyel
J.A.Foster	P.Lee	J Van-Leerzem
W.K.V.Gale	J.C.Martin	N.Williams
H.E.Green	C.Pealing	

Information has also been provided by Baguley- Drewry Ltd, and the NCB.
Records have been consulted at the Public Records Offices at Chancery Lane and Kew, the Staffordshire Joint Records Office at Lichfield and Stafford, the Warwickshire Record Office, Birmingham Library and Archives, Burton-on-Trent Library, Cannock Library, Tamworth Library, Walsall Archives and Wolverhampton Library and Archives.

The original artwork for the maps was produced by Philip Hindley, with final lettering and production by Map-X Visuals of Kilkhampton.

Material consulted includes specific archive material, maps, newspapers and periodicals:

Aris's Birmingham Gazette
Birmingham Post
The Blackcountryman
(Magazine of the Black Country Society)
Burton Daily Mail
Cannock Advertiser
Colliery Guardian
The Engineer
Engineering
Essington Parish News
Gas Journal
Hunt's Mineral Statistics
Journal of the Iron and Steel Institute
Mining Journal
Proceedings of the Institute of Mechanical Engineers
Railway Magazine
Tamworth Herald
Transactions of the Newcomen Society
Walsall Advertiser
Walsall Free Press
Walsall Observer
Wolverhampton Chronicle
Wolverhampton Herald

Also the following publications :

The Canals of the West Midlands,
    Charles Hadfield
The Canals of the East Midlands,
    Charles Hadfield
Guide to the Iron Trade 1873, Griffiths
A History of the Cannock Chase Colliery Co,
    Roger Francis
Noted Breweries of Great Britain & Ireland,
    Alfred Barnard
The Other Sixty Miles,
    Richard Chester-Browne
The Railway to Wombourn, Ned Williams
Track Layout Diagrams of the GWR and BR (WR),   R.A.Cooke
The Midland Railway - A Chronology
    J.V.Gough (R&CHS, 1989)

Ray Shill
100 Frederick Road
Stechford
Birmingham B33 8AE   June 1993

EXPLANATORY NOTES

GAUGE
The gauge of the railway is given at the head of the locomotive list. If the gauge is uncertain, then this is stated. Metric measurements are used where the equipment was designed to these units.

GRID REFERENCE
An indexed six-figure grid reference is given in the text, where known, to indicate the location of salient features such as the loco shed, main works or pit shafts.

NUMBER, NAME
A number or name formerly carried is shown in brackets (   ), if it is an unofficial name or number then inverted commas are used "   "

TYPE
The Whyte system of wheel classification is used wherever possible, but when the driving wheels are not connected by outside rods but by chains or motors they are shown as 4w, 6w, 8w etc. The following abbreviations are used :-

T	Side Tank or similar. The tanks are invariably fastened to the frame.
CT	Crane Tank, a tank locomotive equipped with load lifting apparatus.
PT	Pannier Tank, a special type of side tank where the tanks are not fastened to the frame.
ST	Saddle Tank, a tank which covers the boiler top. This type, usually round, also includes the 'Box' and 'Ogee' versions popular amongst certain manufacturers during the nineteenth century.
IST	Inverted Saddle Tank, a design where the tank passed under the boiler.
WT	Well Tank, a tank located between the frames below the level of the boiler.
VB	Vertical boilered locomotive.
ad	Axle drive. Refers to 'roadrailer', trapped rail systems.
CA	Compressed air locomotive.
DM	Diesel locomotive; mechanical transmission.
DMF	Diesel locomotive; mechanical transmission, flameproof for working underground.
DE	Diesel locomotive; electric transmission.
DH	Diesel locomotive; hydraulic transmission.
PM	Petrol or Paraffin locomotive; mechanical transmission
PE	Petrol or Paraffin locomotive; electric transmission
R	Railcar, a self-propelled vehicle primarily designed to carry passengers.
BE	Battery powered electric locomotive.
BEF	Battery powered electric locomotive, flameproof for underground working.
CE	Conduit powered electric locomotive.
RE	Third rail powered electric locomotive.
WE	Overhead wire powered electric locomotive.
F	Fireless steam locomotive

CYLINDER POSITION

IC	Inside cylinders
OC	Outside cylinders
3C	Three cylinders
4C	Four cylinders
VC	Vertical cylinders
G	Geared transmission (used with IC, OC or VC)

In the case of non-steam locomotives this column is left blank.

MAKERS

Abbreviations used to denote makers are listed in section 8.

MAKERS NUMBER AND DATE

The first column shows the works number, the second shows the date which appeared on the works plate, or the date the engine was completed if no date appeared on the plate.

REBUILDING DETAILS

These are denoted by the abbreviation 'Reb'. They usually record significant alterations to the locomotive.

SOURCE OF LOCOMOTIVE

'New' indicates that a locomotive was delivered from the makers to a location. A bracketed letter indicates that a locomotive was transferred to this location or its temporary transfer away. Full details, including the date of arrival, appear in the footnotes below.

DISPOSAL OF LOCOMOTIVES

A locomotive transferred to a another location is shown by a bracketed number and footnote. Temporary transfers are also indicated in this column where the locomotive no longer remains at the location in question.

In other cases the following abbreviations are used:

OOU	Locomotive noted to be permanently out of use. A date when this was noted is usually given.
Dere	Locomotive noted to be derelict and no longer capable of being used.
Dsm	Locomotive noted in a dismantled state.
Scr	Locomotive broken up for scrap on date shown
s/s	Locomotive sold or scrapped, disposal unknown.

Many sales of locomotives have been effected through dealers and contractors and details are given where known. If the dealer's name is followed by a location e.g. Abelson, Sheldon, it is to be understood that the locomotive went to the Sheldon depot before resale. If no location is given, the loco either went direct to its new owner or else definite information regarding this point is lacking. If a direct transfer is known to have been effected by a dealer, the word 'per' is used.

GENERAL ABBREVIATIONS

c	circa; i.e. about the date quoted
orig	originally
q.v.	mentioned elsewhere in the text
Reb	rebuilt
retn	returned

Geographical abbreviations are found in the text (Section 7) to indicate the course of non-locomotive worked lines:
N- North, E - East, S- South, W - West, SW - South West etc.

Other abbreviations are used for certain firms, canal and railway companies.

Abelson	Abelson & Co (Engineers) Ltd, Sheldon, Birmingham.
BC	British Coal
BCN	Birmingham Canal Navigation.
BR	British Railways.
BR (LMR)	British Railways, London Midland Region.
BR (WR)	British Railways, Western Region.
Cashmore	John Cashmore Ltd, Great Bridge.
CCWR	Cannock Chase & Wolverhampton Railway.
CRC	Cannock & Rugeley Colliery Co Ltd.
CW	Cowlishaw, Walker & Co Ltd, Biddulph, Stoke-on-Trent
EMGB	East Midlands Gas Board.
GCR	Great Central Railway
GER	Great Eastern Railway.
GJR	Grand Junction Railway.
GNR	Great Northern Railway.
GWR	Great Western Railway.
LBSCR	London, Brighton & South Coast Railway
LEW	NCB Lambton Engine Works, Philadelphia, Co Durham.
LMS	London Midland & Scottish Railway.
LNWR	London & North Western Railway.
MPD	Motive Power Depot.
MOM	Ministry of Munitions.
MOS	Ministry of Supply
NCB	National Coal Board.
NCBOE	National Coal Board Opencast Executive.
NER	North Eastern Railway.
NSR	North Staffordshire Railway.
PLH	Petrol Locomotive Hirers.
RHLM	Robert Heath & Low Moor Ltd
ROF	Royal Ordnance Factory.
RPS	Railway Preservation Society.
SSR	South Staffordshire Railway.
WD	War Department.
WDLR	War Department Light Railways.
WMGB	West Midlands Gas Board.
WMR	West Midland Railway.

DOUBTFUL INFORMATION

Information known to be of doubtful nature is printed in brackets with a question mark. Thus 'AB (288?) 1885' means that the locomotive was certainly an Andrew Barclay of 1885 vintage and may have been the works number 288.

SPECIFIC INFORMATION CONTAINED IN THE TEXT

This Handbook deals with an area where agriculture is mixed with industry. South Staffordshire is also an area rich in industrial heritage. It is therefore important to define certain terms and expressions referred to in the text.

Coal Industry

The word **colliery** is a term frequently used in the text. It refers either to a collection of separate coal pits or a mine with extensive plant designed to sort the coal. In the first case individual pits may be sunk to coal measures at different depths and only a limited area of coal worked underground. In medieval times the earliest of these pits consisted simply of a shallow shaft to gain the coal at the base. These are often called Bell Pits and many were sunk in the Cannock district. Later, as techniques evolved, men were able to tunnel underground to extract more of the mineral. In the second case, more characteristic of modern mines, the underground workings might cover several miles and many acres of coal could be worked. Most of the mines described in this book fall into this category. They often had extensive railway systems both above and below ground.

Iron Industry

The manufacture of iron was originally a backwoods process in which iron ore was smelted with charcoal in a **Bloomery**. These, often little more than open hearths, were frequently located on windswept sites. On Cannock Chase there were several where the wood from the forest provided the charcoal and nearby mines of ironstone provided the ore. By the sixteenth century these bloomeries had given way to **furnaces** where the ironmaking process was more efficient. The blast for the fire was provided by water powered bellows and so the ironmaking industry moved from hillside to riverside. The water wheel was an adaptable machine which in addition to producing a blast powered the hammers which worked the crude iron or split iron rods for nail manufacture. At suitable locations water powered mills were built beside the county's rivers and streams. These included the Penk, Tame and Stour, with several on the latter near Kinver. During the eighteenth century the power of steam was harnessed and the iron making process moved again, this time to the coalfields of the area.
The term **forge** refers to the building where the iron is heated in a furnace and then wrought or worked by steam or water power.

Transport

South Staffordshire possessed an intricate system of canals and railways which were constructed for a variety of purposes. The transport of coal was an important factor in their development as well as the conveyance of bricks, beer, iron, stone and general merchandise. The canals handled both long and short distance work. There were the day boats which travelled only a few miles and cabin boats with attendant crews which journeyed longer distances.

The West Midlands canals were a link in a chain which stretched from Lancashire to London. They were plied by long distance carriers and local firms alike. Though the canals lost trade to the railways they still retained a local business in the West Midlands. In fact, this local trade increased through railway patronage and new lines of canal came to be constructed. The Cannock Extension Canal was not completed until 1863 when the Cannock coalfield was developed, and it remained an important carrier of coal until the early 1960's

The Birmingham Canal Navigation comprised several canals mostly built by themselves but incorporating two others viz the Dudley Canal (amalgamated 27/07/1846) and Wyrley and Essington (amalgamated 4/1840). There were ten different levels of the BCN and 66 separate branches. Wherever possible references are made in the text to identify the canal which served an individual industrial location.

In addition to the BCN, the following other canals are mentioned in the book:

    Birmingham & Liverpool Jct   Staffordshire and Worcestershire
    Bond End   Trent & Mersey
    Coventry   Wyrley & Essington

The River Trent was also a navigable waterway as far as Burton-upon-Trent. Improvements, opened in 1712, were made to the river under the Trent Navigation Act of 1699. There were however long periods when vessels could not reach Burton either because of drought or flood.

The Bond End Canal

During 1774 the lessees of the Trent Navigation constructed a canal from the head of navigation at Bond End to a wharf beside the Trent & Mersey Canal at Shobnall, where goods could be transhipped. Traffic on the canal was worked by the Burton Boat Company who kept three vessels specially for the service. At that time the Burton Boat Company was an extensive carrier on the inland waterways.
In 1795 the Bond End Canal was connected to the Trent & Mersey Canal and other craft came to use it but it saw little use after 1870 and closed about 1872. The canal bed was filled in and utilised by the Midland Railway to form the Bond End and Shobnall branches. Shobnall Wharf beside the Trent & mersey Canal was retained as a railway transhipment basin.

There are frequent references in the text to the different railways which served South Staffordshire's industry. Their respective histories do not lie within the scope of this book, but it is important to note their changes of name and ownership. The industrial railways of the area were served by two main railway companies, the Midland Railway and the LNWR. They handled the bulk of the industrial traffic. But they, in turn, comprised several constituent companies, the relevant ones for this area being:

	Date Rly Incorporated	Date Rly completed	Date of vesting / leasing
London and North Western Railway:	16/07/1846	...	...
Cannock Chase Railway	15/05/1860	/1863	28/07/1863
Cannock Mineral Railway	14/08/1855	07/11/1859	07/11/1859
Grand Junction Railway	06/05/1833	04/07/1837	16/07/1846
South Staffordshire Railway	03/08/1846	01/05/1850	20/05/1861

Midland Railway :	10/05/1844	...	...
Birmingham & Derby Junction Railway	19/05/1836	10/02/1842	10/05/1844
Wolverhampton & Walsall Railway	29/06/1865	01/11/1872	01/07/1876
Wolverhampton, Walsall & Midland Junction Rly	06/08/1872	01/07/1879	30/07/1874

The Wolverhampton and Walsall Railway was vested in LNWR 1/7/1875 and worked by it until 31/7/1876.

Great Western Railway	31/08/1835		
Shrewsbury & Birmingham Railway	03/08/1846	12/11/1849*	01/09/1854

* Section Oakengates - Wolverhampton.

Other Railways of note:
North Staffordshire Railway 26/06/1846   11/09/1848#
 # Uttoxeter to Burton Jct (later known as North Stafford Jct)

Great Northern Railway 26/06/1848
 This company reached Burton, opening its own goods depot, on 01/07/1878. Running powers were exercised over the North Staffordshire Railway from Dove Jct.

The Railways Act, 19/8/1921, provided for the reorganisation of the railways through a system of grouping. Under this process the Midland Railway and the LNWR became part of the LMSR. The Act became effective from 1/1/1923. The British Transport Commission took over control of the LMS and GWR from 1/1/1948, their systems becoming part of BR (LMR) and BR (WR) respectively.

Burton on Trent once had a very complicated system of railways. Four companies served the town - the Midland, London & North Western, North Staffordshire and Great Northern Railways. The main line of the Midland Railway - the major operator - passed to the west of the town and several branches were constructed to the important breweries. These branches were worked by Midland Railway engines and brewery company locomotives (which had running powers over them). LNWR locomotives also worked over these branches and had a wharf at Moor Street on the Bond End branch.

Brief details of these branches are as follows:

BOND END BRANCH (Opened 1875)   C
From Wellington Street Junction to the River Trent. At Dale Street Junction there was a branch to Leicester Junction (opened 1884). At Uxbridge Street Junction ther was a branch to Duke St, also known as the Town Branch, and a siding to Wood Street. After a level crossing with the Lichfield Street, the line terminated at Bond End Wharf. Parts of the line were worked by Bass Ratcliff & Gretton's locomotives as well as those of the railway companies.

DALLOW LANE BRANCH   G
An LNWR branch which ran from the junction with the Midland Rly Shobnall Branch and served Allsopp's Shobnall Maltings, Staton's Cement Works and Yeoman, Cherry & Curtis' Brewery before terminating at Stretton Junction (NSR).

DUKE STREET or TOWN BRANCH   **K**
Ran from a junction with the Bond End Branch at Uxbridge Street. At James St Junction it connected with the New Street Branch and then crossed Park Street and New Street. Two crossings at Duke Street, one for Eadie's Brewery and the other for Bass, were worked by one box. Worked by Bass Ratcliff & Gretton's locomotives. Opened 12/1875.

ROBINSONS or NEW STREET BRANCH   **H**
Ran from James Street Junction to Park Street crossing, Robinsons Brewery sidings, Charrington Brewery sidings, Worthingtons Brewery. Worked by Charrington's and Worthington's locos as well as those of main line companies. Opened 26/04/1880.

GUILD STREET   **A**
From Guild Street Junction to the Hay Branch. Sidings to S.Allsopp's New Brewery at Brook St, level crossing Allsopp's Railway, sidings to Bass Ratcliff & Gretton Middle Brewery and loco shed, Guild Street level crossing, Church Croft Junction, High Street level crossing, sidings to Bass/ Allsopp Old Breweries. Worked by S. Allsopp (later Ind Coope Ltd) and Bass, Ratcliff & Gretton locomotives.

HORNINGLOW   **D**
From Horninglow Bridge to Trent and Mersey Canal.

HAY   **B**
From Dixie Sidings (Bass, Ratcliff & Gretton) to The Hay. Hawkins Lane Junction (LNWR), Anderstaff Lane, Salt's maltings and loco shed, Burton Brewery, Salt's Brewery, Allsopp's Old Brewery, Hay sidings, Junction Guild Street branch, Bass's Old Brewery, Worthington's Brewery. Worked by Burton Brewery Co, T.Salt, Worthington and Co.

MOSLEY STREET   **J**
From Burton Station to Ind Coope's brewery in Station Street, opened 13/3/1865.

SAUNDERS (also referred to as **SANDERS** in some sources)   **E**
A short branch beside Anderstaff Lane which served Bass's maltings.

SHOBNALL BRANCH   **F**
(Opened c11/1874, original curve to mainline opened 28/4/1873). Ran from Leicester Junction to Shobnall via Wellington Street Junction, Bond End Branch, Bass Cooperage, Malthouses, LNWR Dallow Lane Branch and Marston, Thompson & Evershed Brewery.

PRIVATE OWNER WAGONS

Apart from the firms listed in section 2, there were other operators of private owner wagons in South Staffordshire. Coal merchants were based in goods yards throughout the region. Many had their own private wagons which they ran from colliery to coal yard. Other operators included firms which retained their own sidings but which were shunted by one of the railway companies mentioned above. It was common to take wagons on a hire purchase agreement and local wagon building firms, such as the Birmingham Railway Carriage & Wagon Co Ltd, provided a great number. In addition to leasing them to local industry these builders also agreed to maintain them for a number of years.

NON-LOCOMOTIVE LINES

Section 7 lists railways and tramways where haulage methods other than by locomotive were employed. They specifically concentrate on those operators who exclusively used this type of railway. Where a firm used a combination of locomotive and non-locomotive worked line, it has been dealt with under section 2.

Non-Locomotive worked lines employed a variety of haulage techniques. Horses were the most common form of power on both standard and narrow gauge railways. Hand propulsion was also common where the distance involved was short. Towards the middle of the nineteenth century, mechanical means of haulage became more common. Ropes or chains were attached to wagons and pulled by a stationary engine. There were three basic versions:

Direct	The engine hauls wagons attached to a rope or chain in one direction and the wagons gravitate back in the other direction taking the rope or chain with them.
Endless	A loop of wire rope or chain is kept in motion by an engine. The rake of wagons is attached by clips above or beneath the body of the leading wagon to the rope or chain and are dragged along the railway.
Main & Tail	A two drum haulage engine is employed with a rope on each drum. One rope is attached to each end of a rake of wagons.

In short sidings and at wagon turntables, capstans worked by hydraulic or electric power were used to assist the movement of traffic.

The area possessed a number of non locomotive worked lines. Some were temporary, others were permanent features. It is not possible to detail all in this book and only those lines longer than 50 yards have been discussed. They were particuarly common at coal mines and blast furnaces. Though not all mines or furnaces used surface railways, canal or road transport might have been chosen instead.

Several contractors have been included in these sections. If a mention is made then that contractor used light railway plant as part of the construction work. This includes railway and some canal building schemes.

Every attempt has been made to include all relevant and major railways and tramways in this part. If there are any omissions, the Industrial Railway Society would gratefully appreciate the information.

Brickworks are generally not included, as their lines tend to be rather short and in several cases there existed only an inclined plane to haul the wagon out of the marl hole.

SECTION 1

KEY MAP
MAP LOCATION INDEX
MAPS SA to SK

ELECTRICITY
keeps the wheels of industry turning

Increased Efficiency—through its great adaptability to all purposes. Instantly available service at the touch of a switch.

Reduced Cost—because it can be switched on or off as required or controlled automatically to meet varying requirements.

Labour Saving—because there is no preparation. Electricity offers—The best and most convenient light, the cleanest form of heat, the most adaptable system of motive power.

Electricity ensures Efficiency & Economy

Whether your requirements be
POWER - LIGHT - HEAT
for works, shop or the home, we shall be glad to advise and quote entirely without obligation.

CORPORATION ELECTRICITY DEPT.
BURTON UPON TRENT

Telephone—Office: 2762; Showroom: 2745

KEY TO MAP AREAS

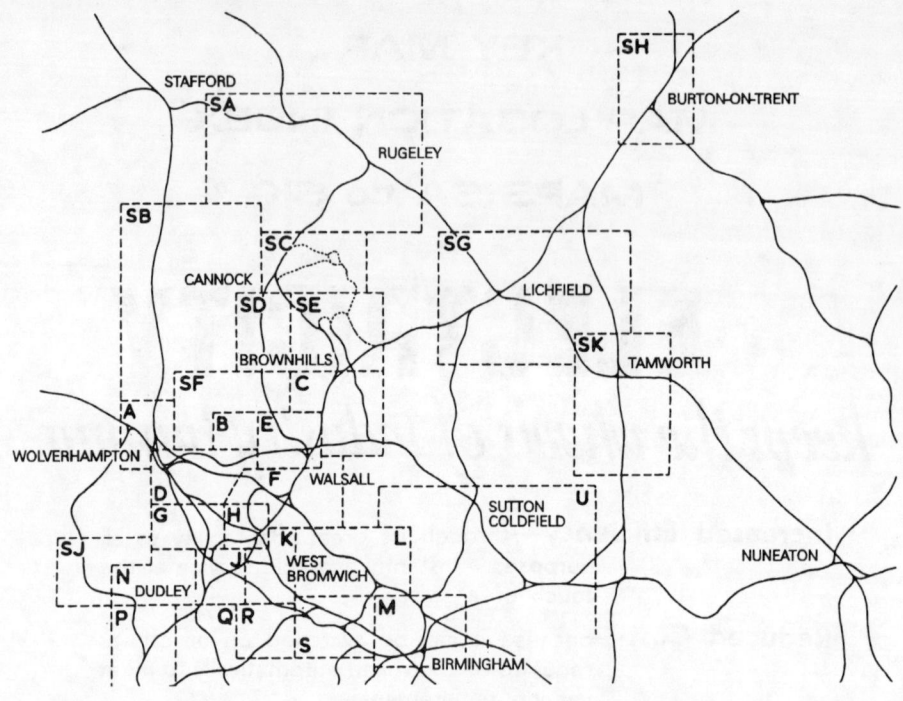

Note that Map Letters A - U refer to West Midlands Handbook.

LEGEND for MAPS :

Railway - Main line.

Railway - Industrial Standard gauge.

Railway - Industrial Narrow gauge.

Canal.

MAP LOCATION INDEX

MAP SA   Rugeley

1. Brereton Colliery Tramway, Earl of Shrewsbury and Talbot.
2. Brereton Colliery Railway, Earl of Shrewsbury and Talbot.
3. Hayes Colliery Tramway, Marquis of Anglesey.
4. Fair Oak Colliery, Fair Oak Colliery Co Ltd.
5. Milford Gravel Pits, Sir Alfred McAlpine.
6. Cannock Chase Military Railway, War Department.
7. Lea Hall Colliery, NCB.
10. Belfast Colliery, Earl of Shrewsbury and Talbot.
11. Brereton (orig. Brick Kiln) Colliery, Earl of Shrewsbury and Talbot /Brereton Collieries Ltd/NCB.
12. Coppice Colliery, Earl of Shrewsbury and Talbot.
13. Hayes Colliery, Earl of Shrewsbury and Talbot.
14. Old Engine Colliery, Earl of Shrewsbury and Talbot.

MAP SB   Penkridge

1. Four Ashes Works, Croda Synthetic Chemicals Ltd.
2. Littleton Colliery, The Littleton Collieries Ltd/NCB.
3. Paradise Factory, ROF.
4. Littleton Colliery Rly.
5. Village Foundry, John Smith.

MAP SC   Hednesford

1. Littleworth Tramway, Birmingham Canal Navigation.
2. LNWR, Cannock Chase Railway.
3. Cannock Wood Colliery, Cannock & Rugeley Coll Co Ltd/NCB.
4. Valley Colliery, Cannock & Rugeley Coll Co Ltd/NCB.
5. Wimblebury Colliery, Cannock & Wimblebury Coll Co Ltd/NCB.
6. Cannock Chase No 6 Colliery, Cannock Chase Colliery Co Ltd.
7. Cannock Chase No 7 Colliery, Cannock Chase Colliery Co Ltd/NCB.
8. Cannock Chase No 8 Colliery, Cannock Chase Colliery Co Ltd/NCB.
9. Cannock Chase No.9 Colliery, Cannock Chase Colliery Co Ltd/NCB.
10. Cannock Chase No.10 Colliery, Cannock Chase Colliery Co Ltd.
11. LNWR, Littleworth Extension Rly.
12. West Cannock No.1 Colliery, West Cannock Colliery Co Ltd/NCB.
13. West Cannock No.2 Colliery (original), West Cannock Colliery Co Ltd.
14. West Cannock No.2 (orig. No.4) Colliery, West Cannock Colliery Co Ltd/NCB.
15. West Cannock No.3 Colliery, West Cannock Colliery Co Ltd.
16. West Cannock No.5 Colliery, West Cannock Colliery Co Ltd/NCB.
17. Hednesford Depot, Railway Preservation Society
18. East Cannock Colliery, East Cannock Colliery Co Ltd/NCB.
19. Rawnsley Shed, Cannock & Rugeley Colliery Co Ltd/NCB.

MAP SD Cannock and Wyrley

1. Great Wyrley No.3 (later Nook & Wyrley) Colliery, Great Wyrley Colliery Co Ltd/ Nook & Wyrley Colliery Co Ltd/NCB.
2. Mid Cannock Colliery, W.Harrison Ltd/NCB.
3. Cheslyn Hay Tileries, Haunchwood Lewis Tileries Ltd.
4. Old Coppice (later Hawkins) Colliery, T.A Hawkins & Son/NCB.
5. Nook No.2 Colliery, Great Wyrley Coll Co Ltd/ Nook & Wyrley Colliery Co Ltd.
30. Cannock and Leacroft Colliery, Cannock & Leacroft Colliery Co Ltd/NCB.
31. Churchbridge Edge Toolworks, W.Gilpin Senior & Co Ltd.
32. Longhouse Colliery, Henry Hawkins.
33. Quinton Colliery, Quinton Colliery Co.
34. Cheslyn Hay Tramroad, Staffordshire & Worcestershire Canal.
35. Wyrley Cannock Colliery, Wyrley Cannock Colliery Co Ltd.

MAP SE Brownhills

1. Cannock Chase No 1 Colliery, Cannock Chase Colliery Co Ltd.
2. Cannock Chase No 2 Colliery, Cannock Chase Colliery Co Ltd.
3. Cannock Chase No 3 Colliery, Cannock Chase Colliery Co Ltd/NCB.
4. Cannock Chase No 4 Colliery, Cannock Chase Colliery Co Ltd.
5. Cannock Chase No 5 Colliery, Cannock Chase Colliery Co Ltd.
6. Norton Canes Canal Wharf, Birmingham Canal Navigation.
7. Conduit Colliery No 3, Conduit Colliery Co Ltd./ Littleton Collieries Ltd
8. Conduit Colliery No 4, Conduit Colliery Co Ltd.
9. Coppice Colliery 1-5, Coppice Colliery Co Ltd.
10. Coppice Colliery 6 & 8, Coppice Colliery Co Ltd.
11. Coppice Colliery, Coppice Colliery Co Ltd/NCB.
12. Cathedral Colliery, W.Harrison Ltd.
13. Grove Colliery, W.Harrison Ltd/NCB.
14. Chasewater Railway, Chasewater RPS.
15. Brownhills No.3 (later Wyrley No.3) Colliery, W.Harrison Ltd/NCB.
16. Chasewater Workshops, NCB.
17. Brownhills Colliery, W.Harrison.
18. Wyrley Common Colliery, W.Harrison.
30. Conduit Colliery 1 & 2, Conduit Colliery Co.

MAP SF Essington

1. Essington Wood Colliery, Darlaston Coal & Iron Co Ltd.
2. Essington Farm Colliery, Essington Farm Colliery Co Ltd.
3. Rosemary Tileries, Haunchwood Lewis Tileries Ltd.
4. Holly Bank Colliery, Hilton Main and Hollybank Colliery Co Ltd/NCB.
5. Hollybank Mineral Rly, Hilton Main & Hollybank Colliery Co Ltd/NCB.
6. Sneyd Colliery, Hollybank Colliery Co Ltd/NCB.
7. Hilton Main Colliery, Hilton Main & Hollybank Colliery Co Ltd.
8. Cannock Lodge Colliery, W Lounds.
9. Norton Cannock Colliery, Norton Cannock Colliery Co Ltd.
10. Essington Disposal Point, NCBOE (First site).
11. LNWR/LMS/BR Essington Mineral Railway.
12. Essington Disposal Point, NCBOE (Second site).
30. Essington Collieries Mineral Tramway.
31. Yew Tree Drift Mine, NCB

MAP SG   Lichfield

1 Beacon Park, Lichfield District Council.
2 Elford Workshops, River Trent Catchment Board.
3 Rom River Co Ltd.

MAP SH   Burton on Trent

1 New Brewery, Samuel Allsopp & Sons Ltd
2 Middle Brewery, Bass, Radcliff & Gretton Ltd.
3 Bass Museum
4 Branston Factory, Ministry of Defence./ Crosse & Blackwell Ltd.
5 Burton Brewery, Burton Brewery Co Ltd.
6 Burton Constructional Engineering Co Ltd.
7 Abbey Brewery, Charringtons Ltd.
8 Wood St Maltings, Charringtons Ltd.
9 Station Street Brewery, Ind Coope & Co Ltd.
10 Wellington Works, Lloyds (Burton) Ltd.
11 Shobnall Brewery, Marston, Thompson & Evershed Ltd.
12 Brewery, T.Salt & Co Ltd.
13 Black Eagle Brewery, Truman, Hanbury, Buxton & Co Ltd.
14 Branston Wagon Works, Wagon Repairs Ltd.
15 Brewery, Worthington & Co Ltd.
16 Branston Gravel Pits, Branston Gravels Ltd.
17 Thornewill & Warham Ltd.
18 Baguley Cars Ltd. / Cyclops Engineering Co Ltd.
19 Baguley Drewry Co Ltd.
20 Shobnall Maltings, Bass, Ratcliff & Gretton Ltd.
21 Shobnall Maltings, Samuel Allsopp & Sons Ltd.
22 Dixie Ale Stores and Cask Washing, Bass, Ratcliff & Gretton Ltd.
23 Old Brewery, Bass Ratcliff & Gretton Ltd.
24 Old Brewery, Samuel Allsopp & Sons Ltd.
25 Anderstaff Lane Cooperage, Burton Brewery Co Ltd.
26 Anderstaff Lane Cooperage, Thomas Salt & Co Ltd.
27 Anderstaff Lane Maltings, Bass, Ratcliff & Gretton Ltd.
28 Middle Yard, Bass, Ratcliff & Gretton Ltd.
29 New Brewery, Bass, Ratcliff & Gretton Ltd.
30 Burton Gasworks, Burton Corporation/EMGB.
31 Robinsons Brewery, Union St
32 J Eadie Ltd, Cross St Brewery.
33 Everards Ltd, Trent Brewery.
34 Peter Walker & Sons, Clarence Brewery.
35 Crown Maltings, L & G Meakin Ltd, later Worthington & Co
36 Burton Power Station, Burton Corporation/CEGB
37 Crescent Brewery, Thos Cooper & Co.
38 Midland Joinery Works Ltd.
39 Curzon St Maltings (later Bottling Stores), Ind Coope & Co Ltd.
40 Shobnall Brewery, A.B.Walker, later Peter Walker & Sons' Brewery.
41 Walsitch Maltings, Thomas Salt & Co Ltd.

Locations on the map with an A suffix, e.g. 1A, are the locomotive sheds associated with the corresponding premises.

MAP SH   Burton on Trent (continued)

For clarity, Railway Companies branches are indicated by letters:

- A Guild Street Branch
- B Hay Branch
- C Bond End Branch
- D Horninglow Branch
- E Saunders Branch
- F Shobnall Branch
- G Dallow Lane Branch
- H New Street Branch
- J Mosley Street Branch
- K Duke Street Branch

Map SJ   Wombourn

- 1   Baggeridge Brick & Tile Co Ltd.
- 2   Perry and Co, Contractors Depot
- 3   Baggeridge Colliery. Earl of Dudley.
- 30 Ferro (Great Britain) Ltd, Wombourn Works.

Map SK   Tamworth

- 1   Alders Paper Mills Ltd.
- 2   Glascote Tileries, Gibbs & Canning Ltd.
- 3   Glascote Colliery, Glascote Colliery Co Ltd.
- 4   Amington Colliery, Glascote Colliery Co Ltd.
- 5   Kettlebrook Colliery, Kettlebrook Colliery Co.
- 6   Wilnecote Colliery, Perrens and Harrison
- 7   Tame Valley Colliery, George Skey & Co Ltd.
- 8   Wilnecote Colliery and Brickworks, George Skey & Co Ltd.
- 9   Kettlebrook Colliery Tramroad, Kettlebrook Colliery Co.
- 10 Dumolo's Colliery, Kettlebrook Colliery Co.
- 11 Peel Colliery, George Skey & Co Ltd.
- 30 Hockley Hall Collieries & Chemical Works.

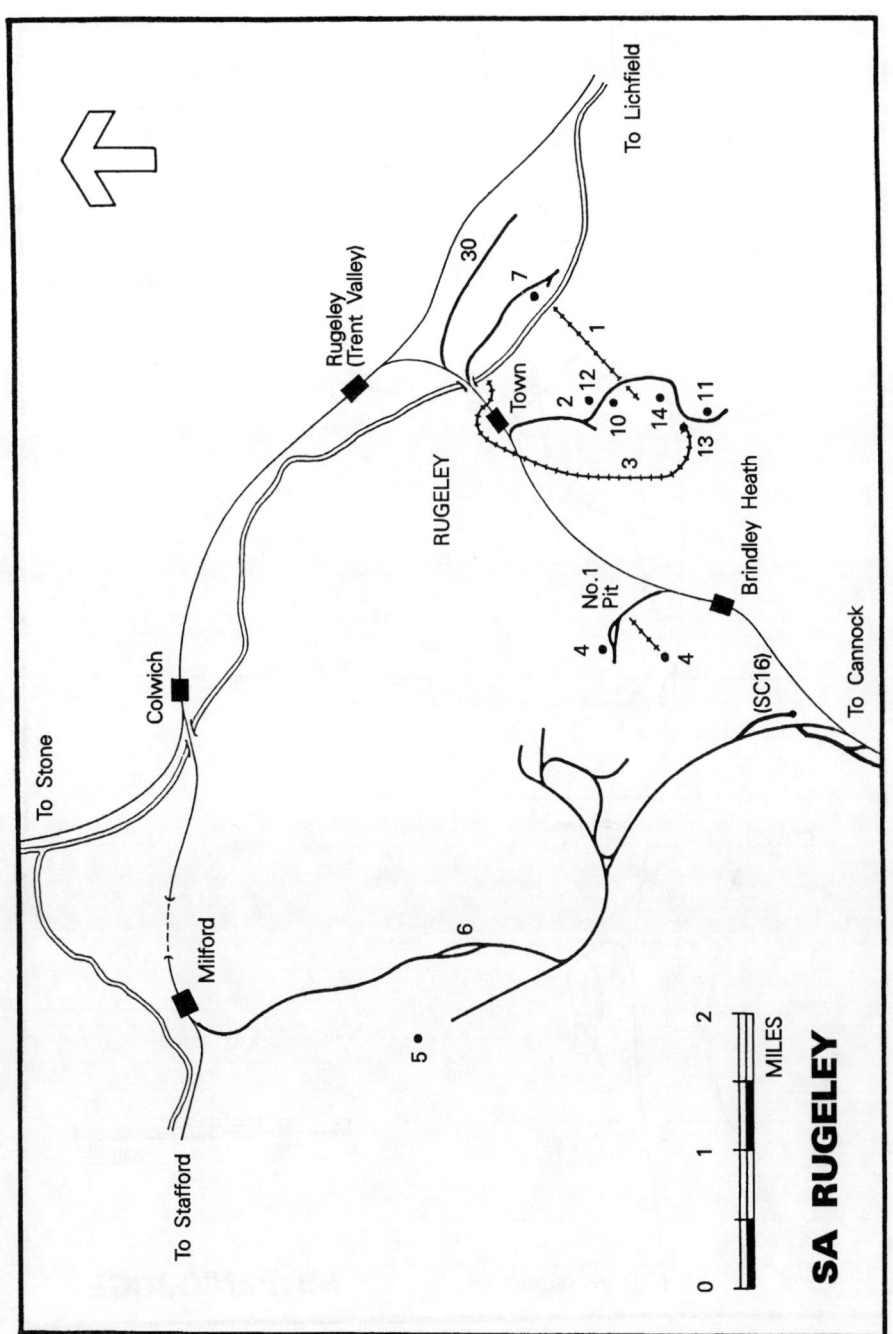

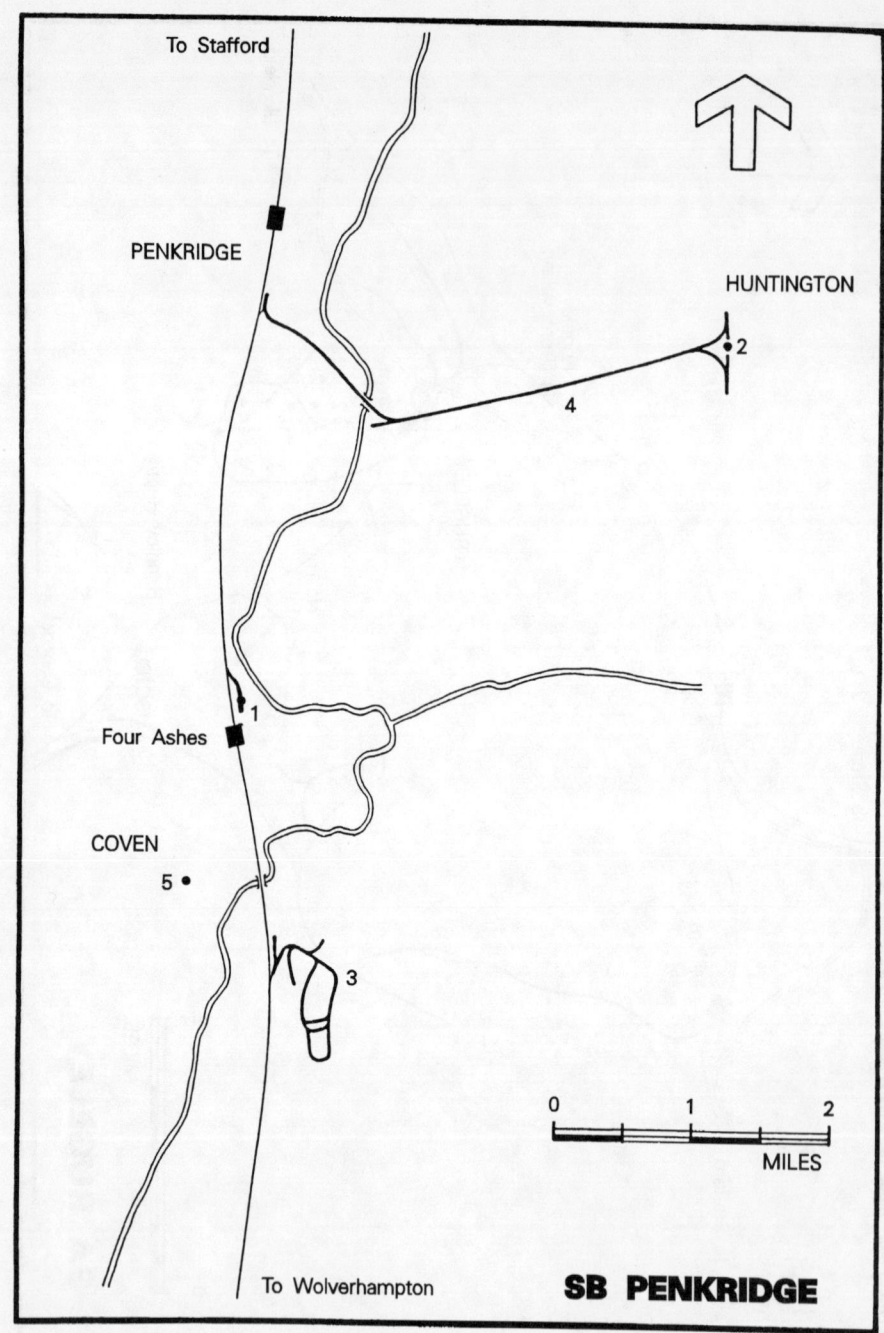

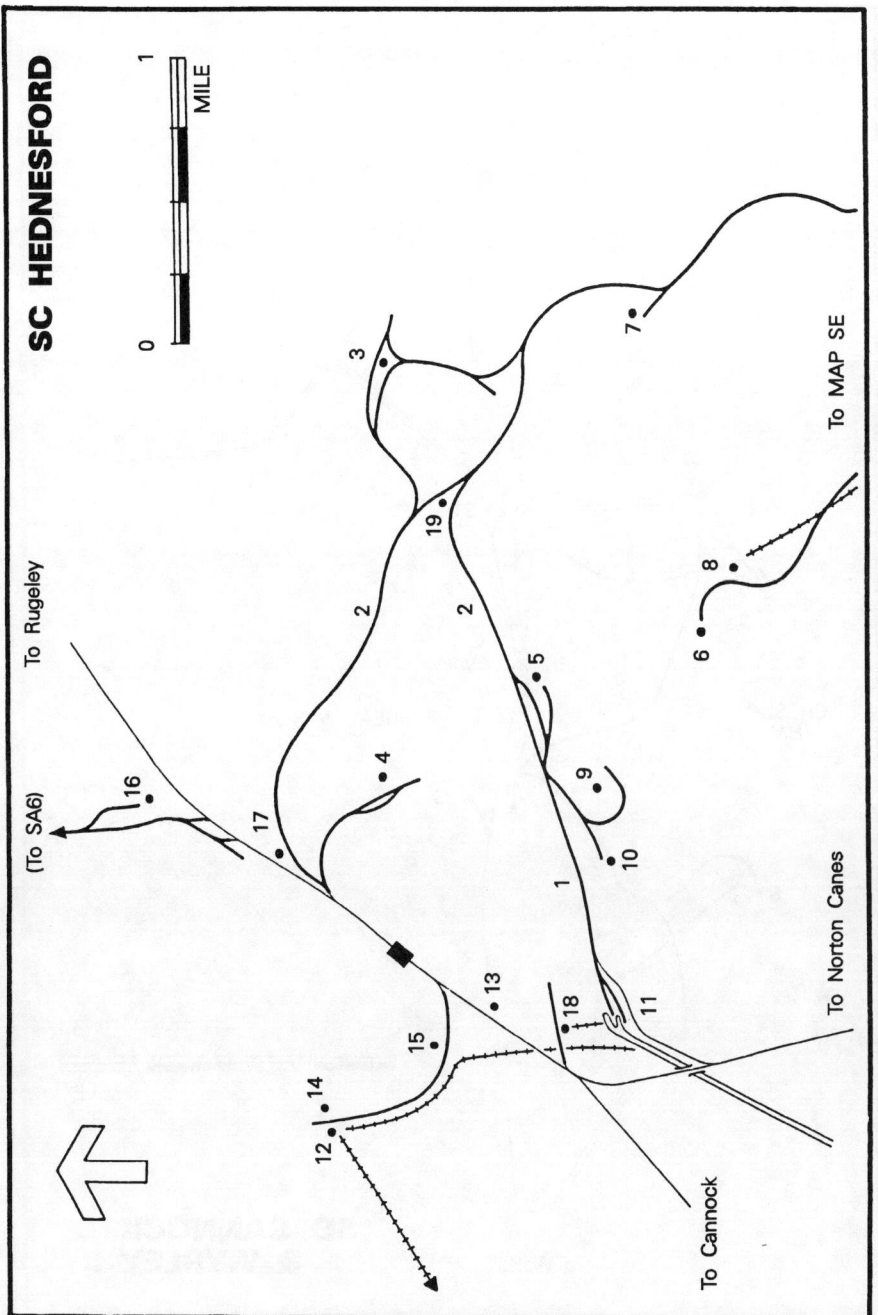

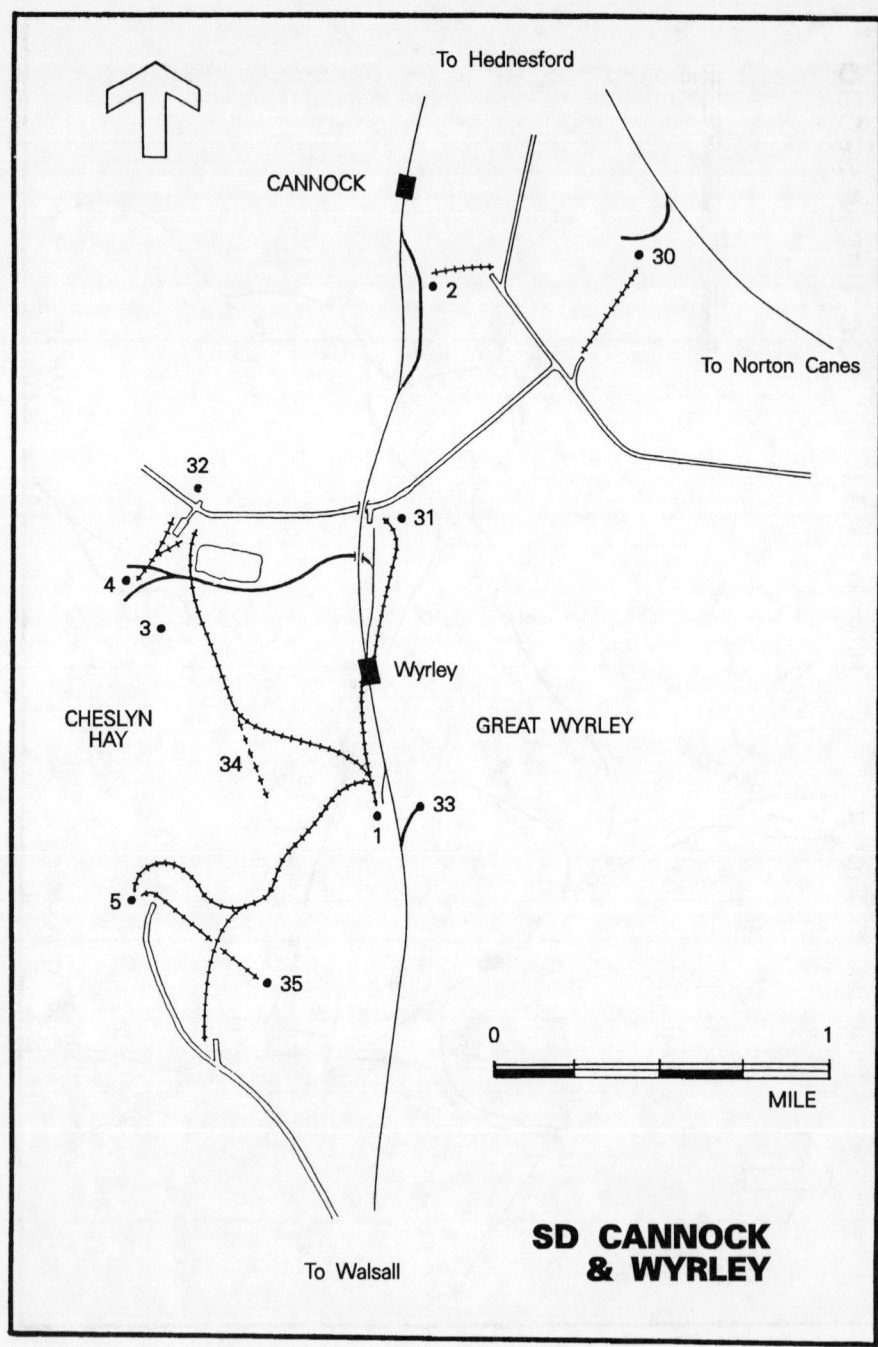

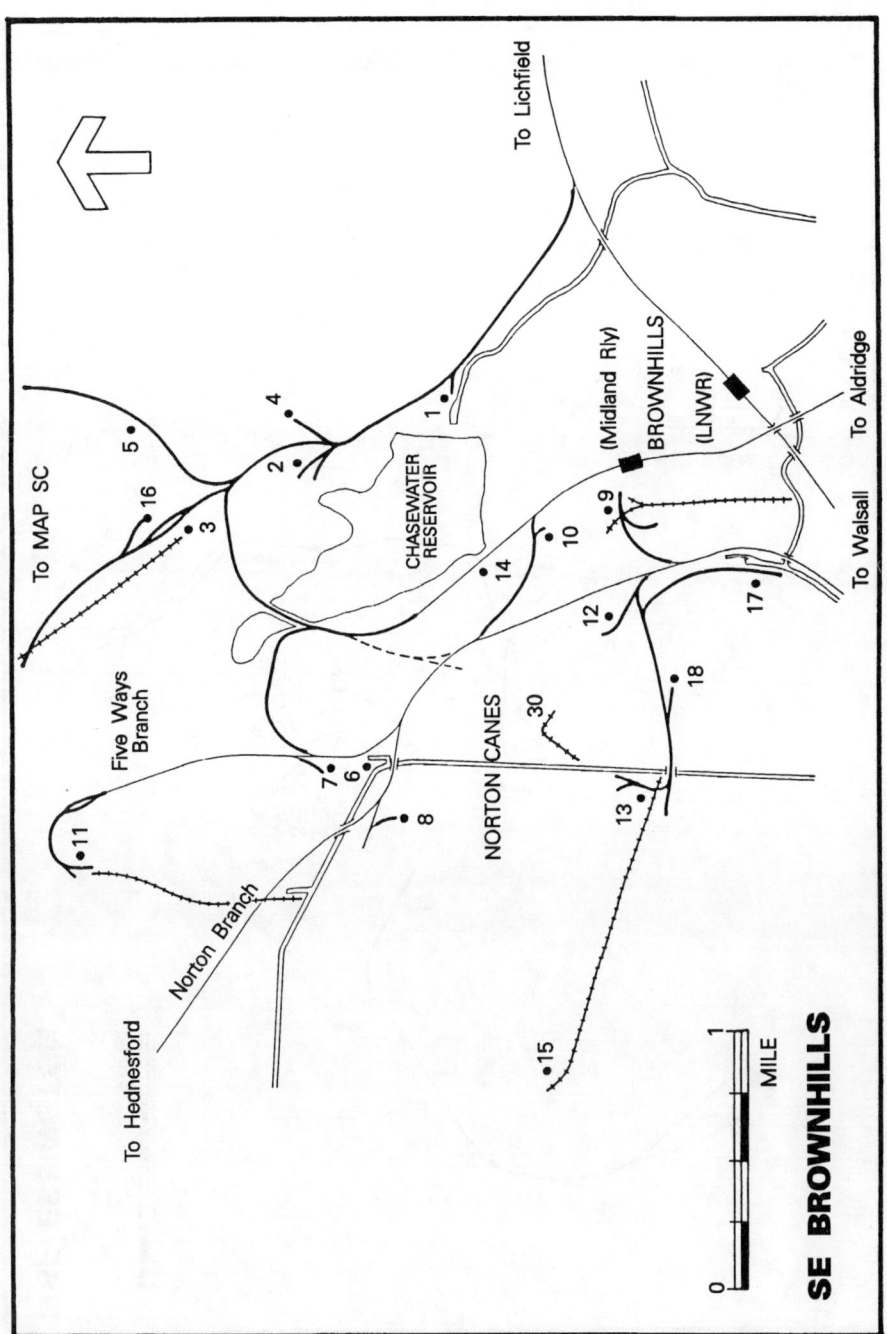

South Staffordshire Handbook Page 29

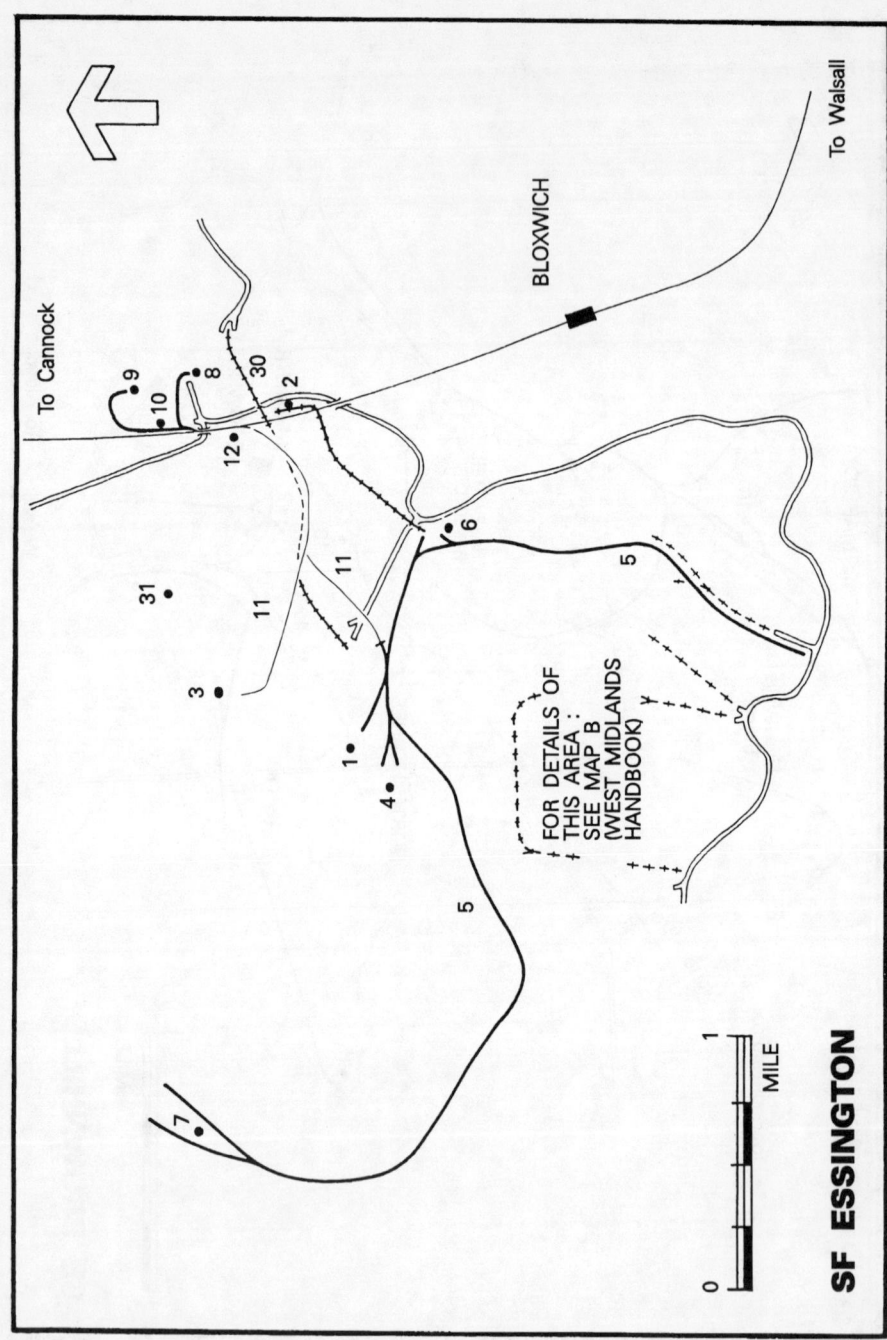

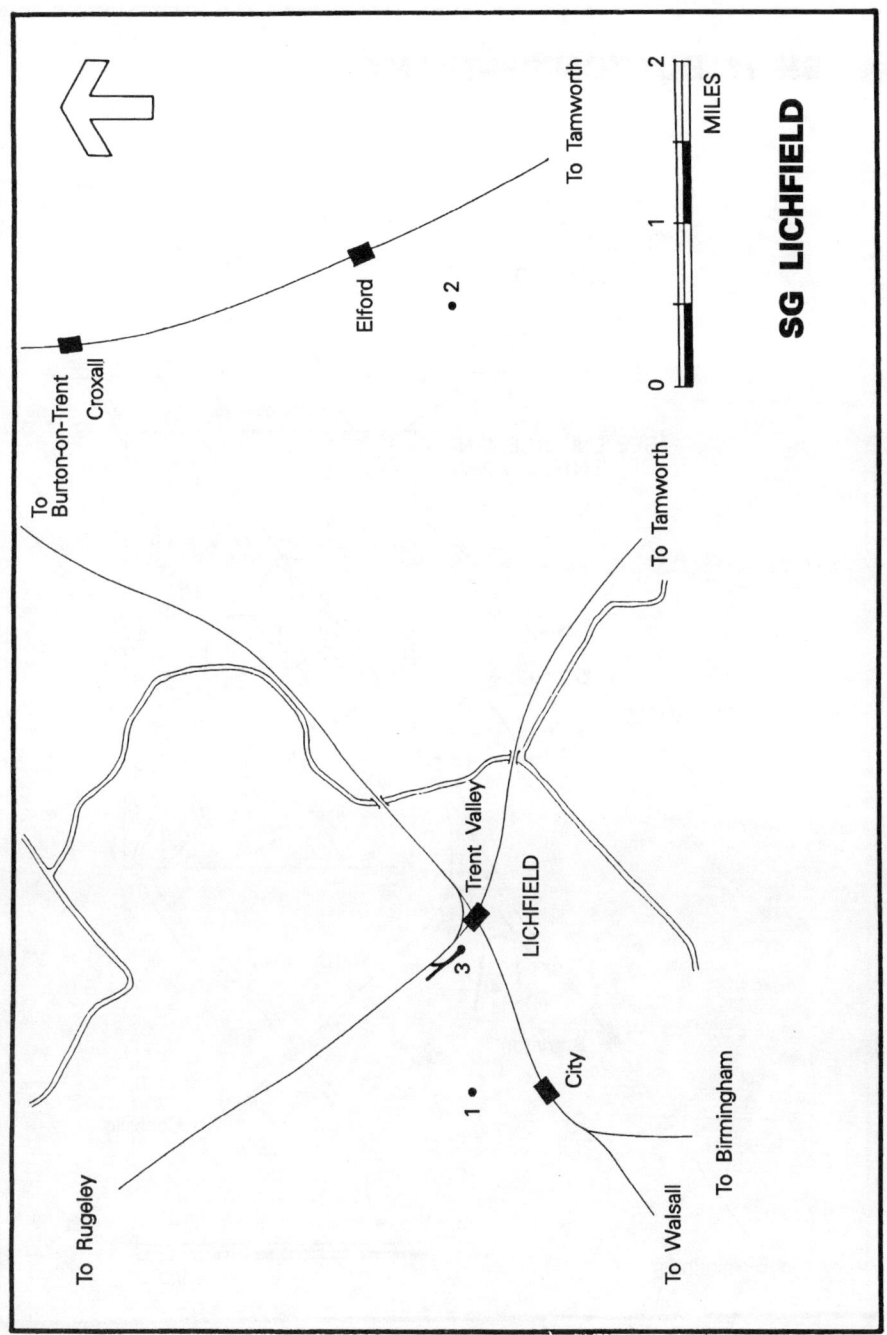

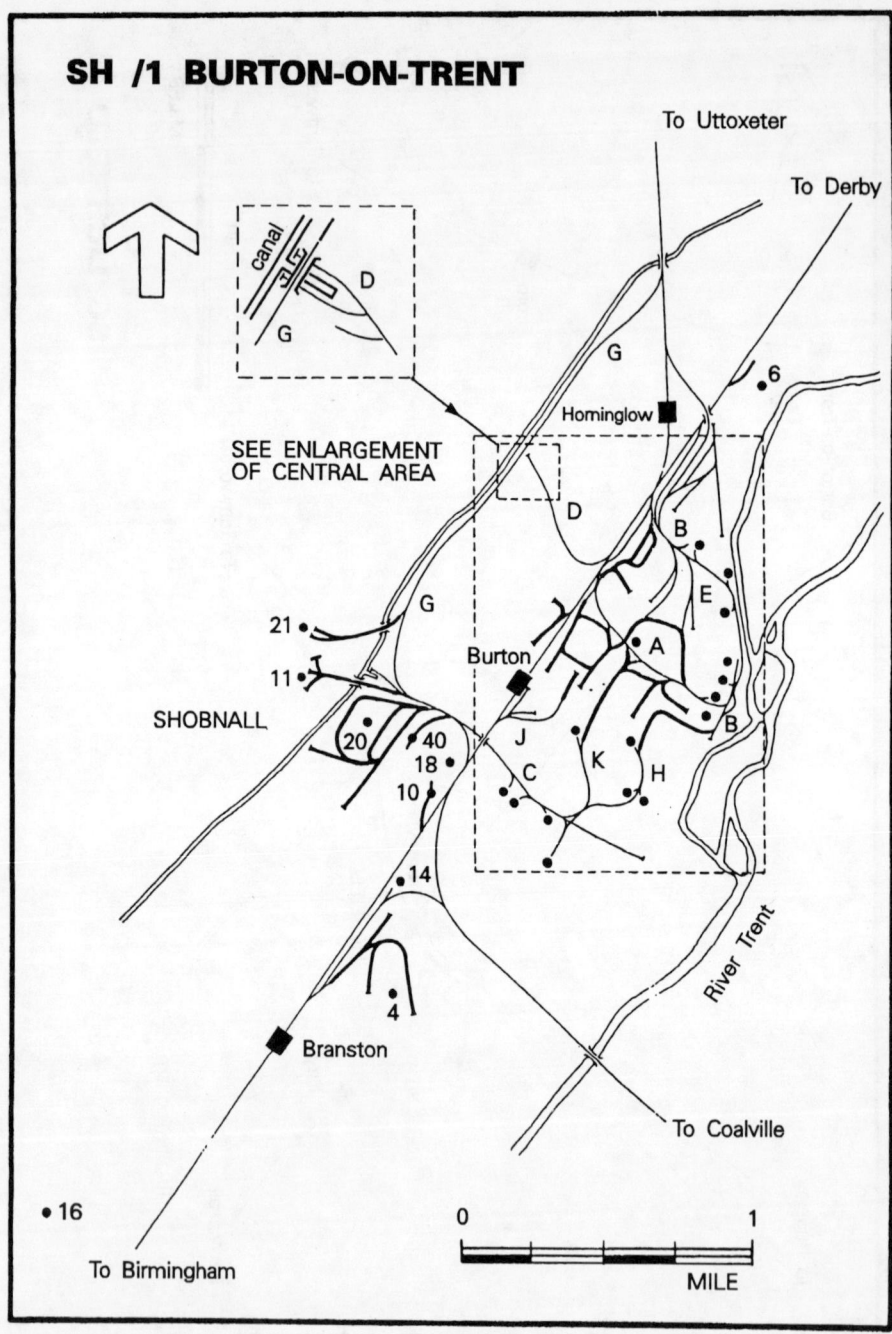

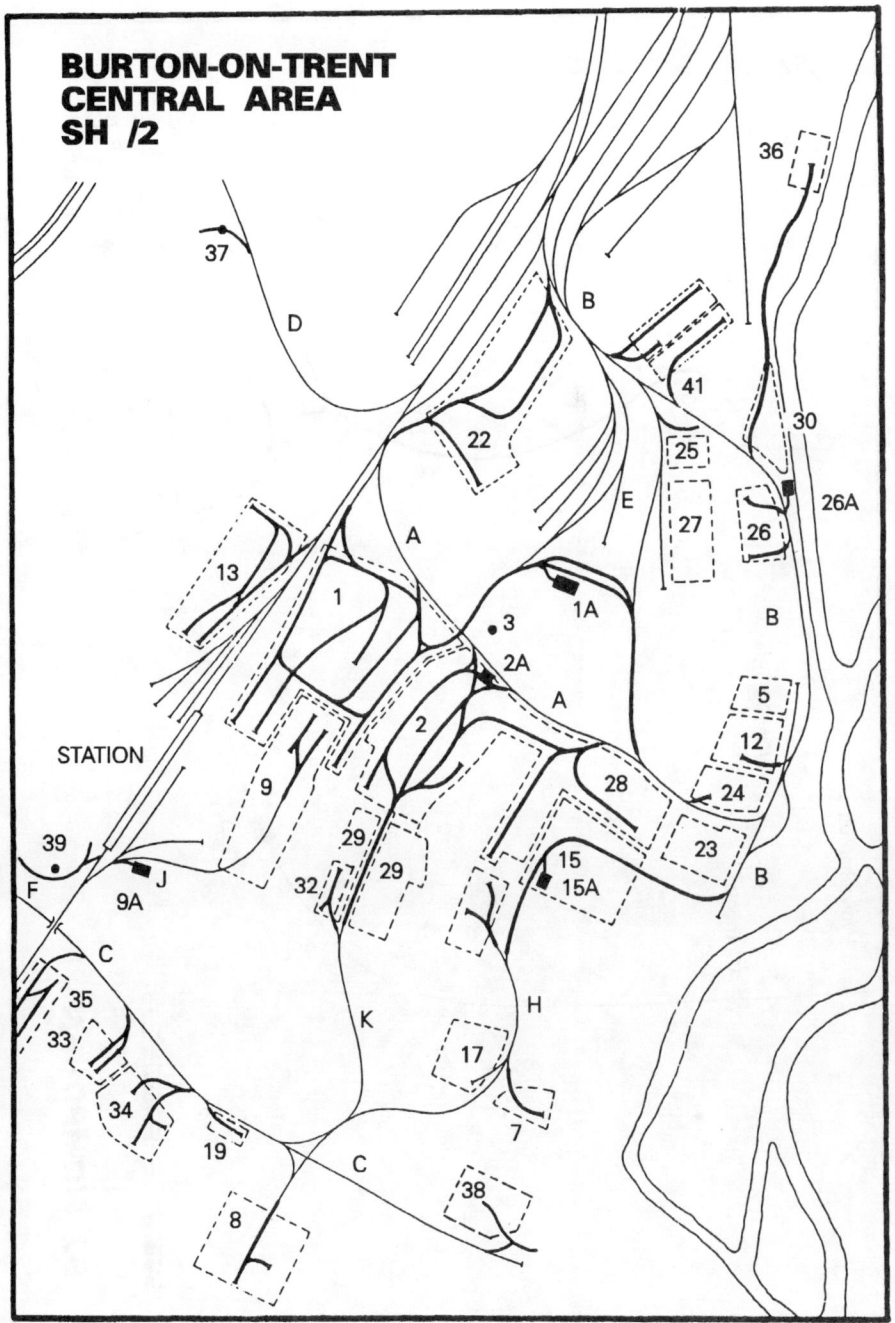

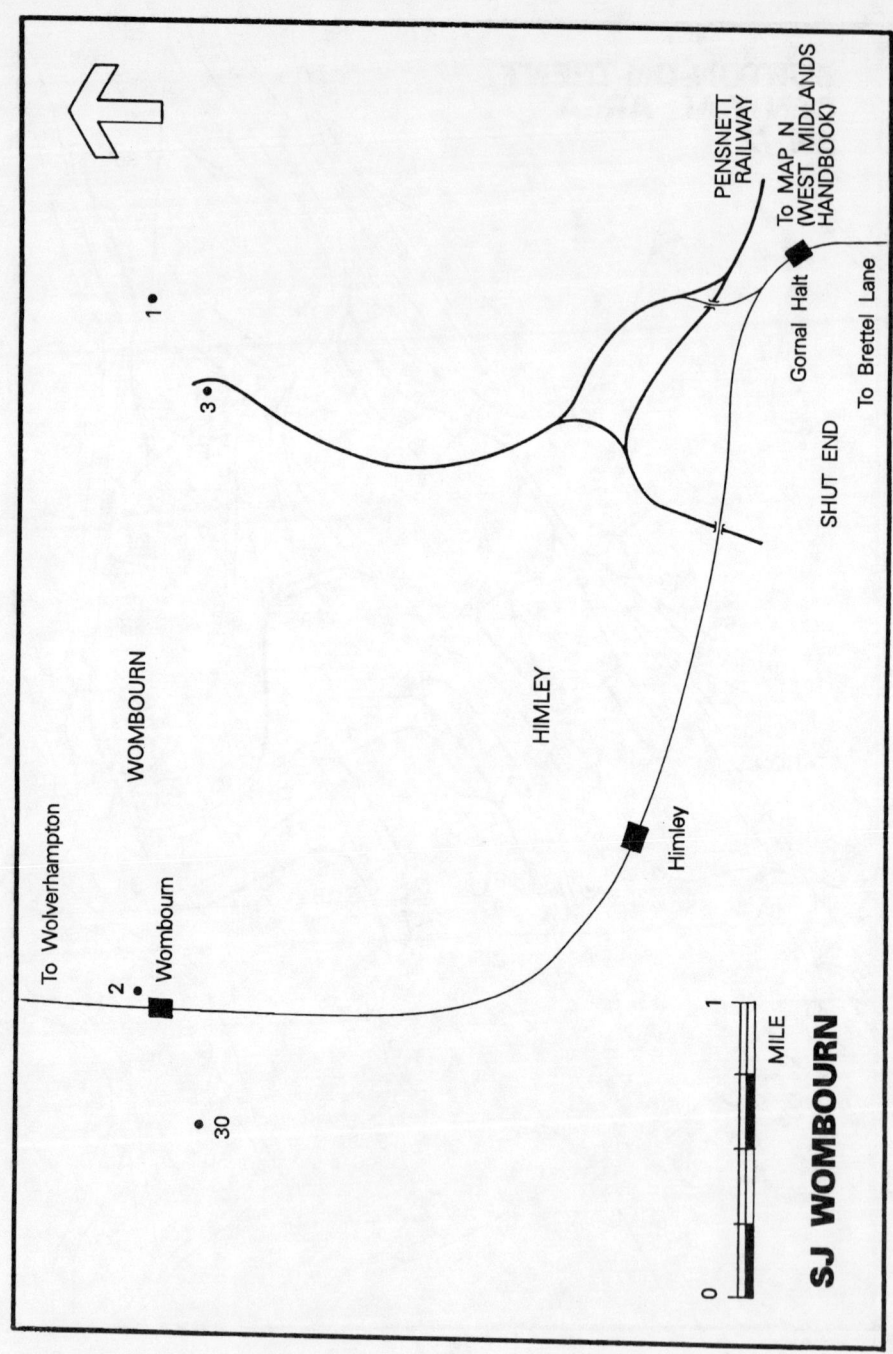

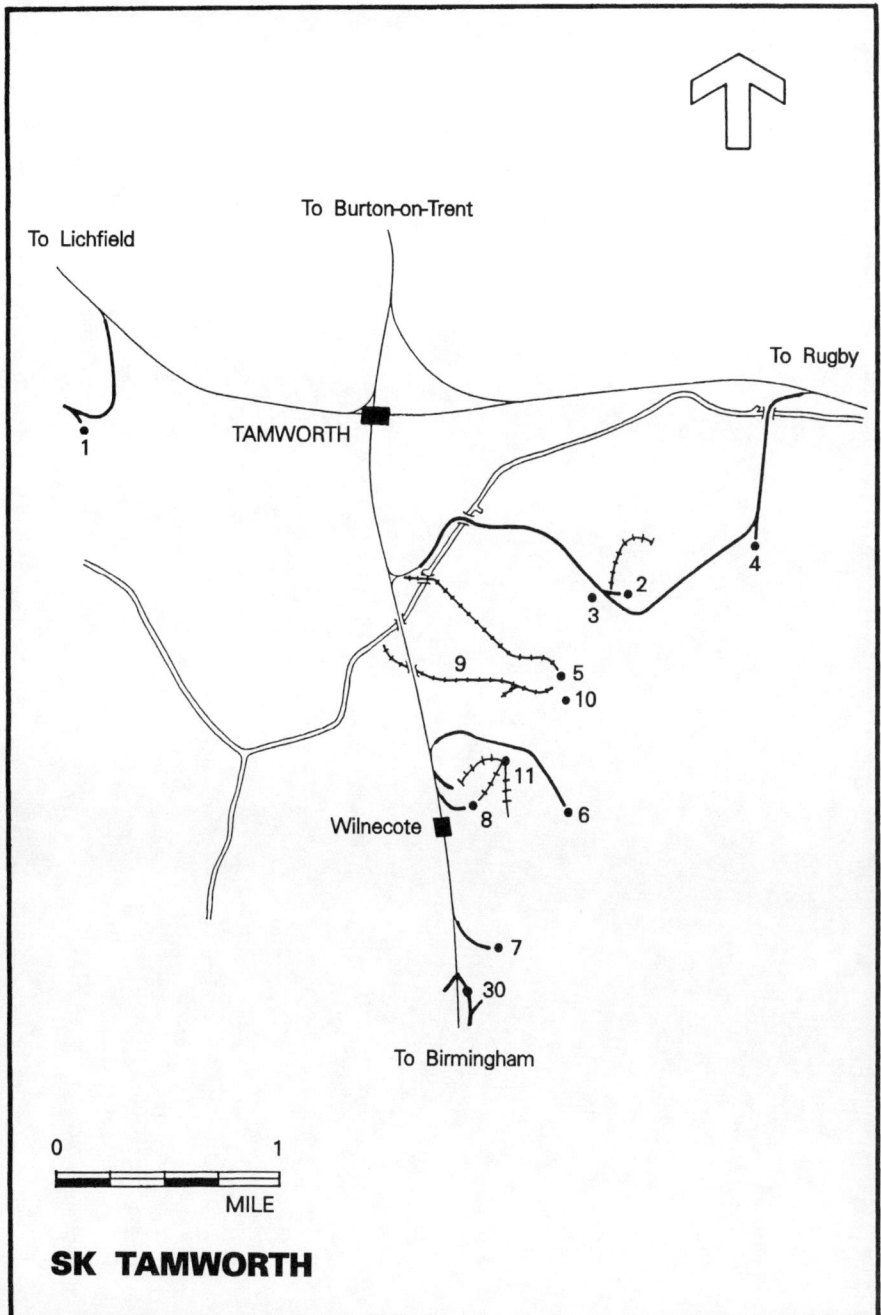

SECTION 2

FIRMS USING LOCOMOTIVES

THORNEWILL & WARHAM,
ENGINEERS, IRON & BRASS FOUNDERS, MILLWRIGHTS, BOILER MAKERS, &c.
BURTON-ON-TRENT, AND DERBY.
London Office :—18, GREAT GEORGE ST., WESTMINSTER.

DIRECT-ACTING PUMPING ENGINE.

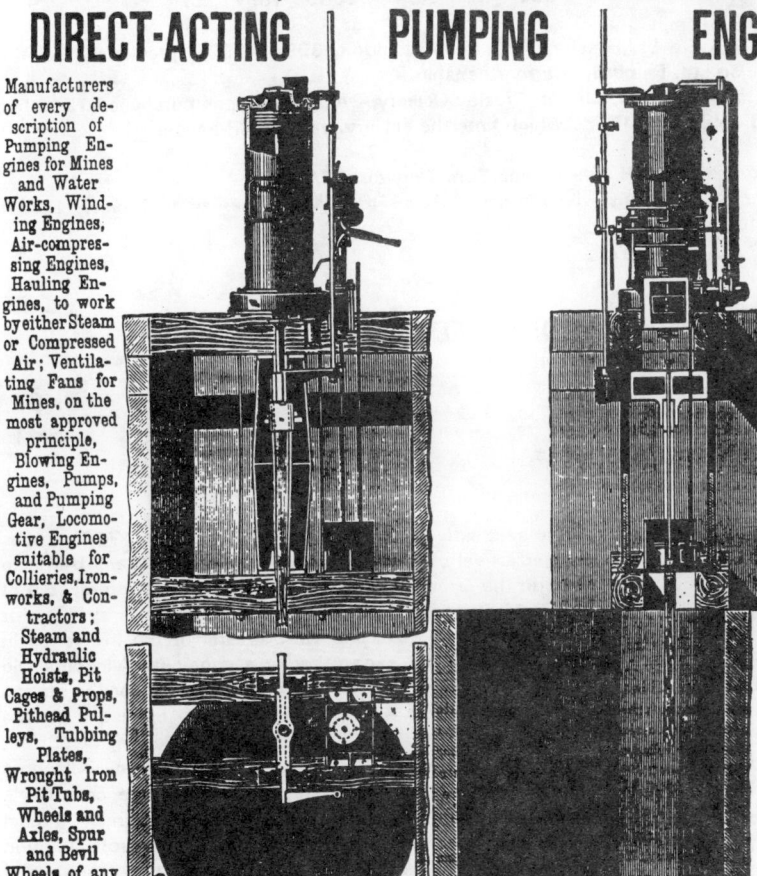

Manufacturers of every description of Pumping Engines for Mines and Water Works, Winding Engines, Air-compressing Engines, Hauling Engines, to work by either Steam or Compressed Air; Ventilating Fans for Mines, on the most approved principle, Blowing Engines, Pumps, and Pumping Gear, Locomotive Engines suitable for Collieries, Ironworks, & Contractors; Steam and Hydraulic Hoists, Pit Cages & Props, Pithead Pulleys, Tubbing Plates, Wrought Iron Pit Tubs, Wheels and Axles, Spur and Bevil Wheels of any diameter,

pitch, or width, made by Patent Machinery; Rotary Steam Pumps for feeding Boilers, suitable for Collieries, Breweries, Chemical Works, &c. Feed-water Heaters, Boilers, Girders, Tanks, and every description of Boiler Mountings, Water Wheels, Clay Mills, Mortar Mills, Crabs, Steam Winches, Castings for Gas Works, Mouthpieces, Purifiers, Hydraulic Mains, Pipes for Gas Holders, &c.; Shafting, Pulleys, Brackets, Plummer Blocks, and Colliery and Brewery Plant in general, with all the latest improvements.

PRICES, &c., FURNISHED ON APPLICATION.

ALDERS (TAMWORTH) LTD
ALDERS PAPER MILLS, Tamworth.　　　　　　　　　　　　　　　　　　　　**SK1**
Alders Paper Mills Ltd until 9/6/1932.

These mills are situated beside the A51 Tamworth to Lichfield road (SK 192045). In 1927 a short branch railway, half a mile long, was laid to connect the factory with the LMS Trent Valley line at Coton Crossing. Rail traffic ceased in October 1967 and track was lifted between 1968 and 1969.

　　Ref:　Industrial Railway Record, No.48, p52-55
　　　　　Industrial Railway Record, No.65, p189-198.

Gauge: 4ft 8½in

		0-6-0ST	OC	AB	1576	1918	(a)	(1)
1340	TROJAN	0-4-0ST	OC	AE	1386	1897	(b)	(2)

(a)　　ex R.H.Neal & Co Ltd, dealer, Park Royal, London, 309, 3/1927; previously Nobel's Explosives Co Ltd, Pembrey, Carmarthenshire.
(b)　　ex Moira Colliery Co Ltd, Church Gresley Colliery, Derbyshire. Loco purchased 1946 but not moved until 12/1947, by which time the colliery was in NCB hands.

(1)　　scrapped on site by C.H.Jenkins and Son, Tamworth 2/1968.
(2)　　sold to Mr J.B.True and despatched 3/1968 to GWR Preservation Society, Didcot Depot, Oxfordshire.

SAMUEL ALLSOPP & SONS LTD
BURTON ON TRENT BREWERIES.
OLD BREWERY　　　　　　　　　　　　　　　　　　　　　　　　　　　　　**SH24**
NEW BREWERY　　　　　　　　　　　　　　　　　　　　　　　　　　　　　**SH1**
Samuel Allsopp & Sons until 2/2/1887.
Originally **Benjamin Wilson**

Benjamin Wilson had a brewery on the east side of High Street (SK 253252) by 1742. It was sold to Samuel Allsopp in 1807. After arrival of the railway in Burton in 1839 any traffic to be sent by rail had to be carted through the streets to the Midland Railway station located at the opposite end of the town. In 1859-60 Parliament approved two Midland Railway Acts for the building of two branch lines to connect the main line with several of the breweries in Burton. These Guild Street and Hay branches met each other at a spot near Allsopp's Old Brewery . In 1860 a new brewery (SK 245234) was built for Samuel Allsopp near the station and railway sidings connected it with the Midland Railway. Allsopp also opened a private railway in 1862 linking the New Brewery with its Old Cooperage across Horninglow Street where connection was made with the LNWR exchange sidings and goods station. Allsopp's locomotive shed (SK 249235) was built at the Old Cooperage premises. The company's private line carried on and crossed Horninglow street again to Church Croft sidings. Samuel Allsopp's railway then rejoined Midland Railway Guild Street branch at Church Croft Junction and ran parallel to it for a short distance. Finally it crossed High Street into Allsopp's Old Brewery. The shape of the railway changed little thereafter and ultimately a fleet of six engines was needed to work it. In 1880 new maltings were built at Shobnall **(SH21)**.

Opened on the 27th October they were connected with the LNWR Dallow Lane branch then under construction. It is recorded that the Midland engine worked traffic as far as Shobnall Junction where a Braddock and Matthews contractors engine took over. When complete in 1881 no doubt a LNWR engine performed this chore.

In June 1934 Messrs Allsopp merged with the firm of Ind Coope whose brewery stood next to the Allsopp New Brewery. The locomotive fleets of the two firms were then combined into one.

For later history see under Ind Coope Ltd.

Gauge: 4ft 8½in

1	LEEDS	0-4-0WT	OC	TW	223	1863	New	(1)
2		0-4-0ST	OC	MW	60	1862	New	(2)
3		0-4-0ST	OC	HCR	75	1866	New	Scr.
		0-4-0ST	OC	HCR	148	1874		
			Reb	HL		1912	New	(3)
4		0-4-0ST	OC	HCR	168	1875	New	Scr /1920.
5		0-4-0ST	OC	HCR	177	1876		
			Reb	HL		1912	New	Scr /1922.
6		0-4-0ST	OC	HCR	178	1876		
			Reb	TW		1901	New	(4)
7		0-4-0ST	OC	HC	647	1903	New	Scr /1926.
4		0-4-0WT	OC	TW	393	1874		
		Reb 0-4-0ST		HE		1900	(a)	Scr /1926.
8		0-4-0ST	OC	AB	1858	1925	New	(4)

(a) ex Bass, Ratcliff & Gretton Ltd, /1916.

(1) to HCR, reb as HCR 187 of 1876 0-4-0ST OC and resold to Holwell Iron Co, Melton Mowbray, Leicestershire.
(2) to North of England Industrial Iron & Coal Co Ltd., Carlton Ironworks, Stillington, Co. Durham.
(3) to Redbourn Hill Iron & Coal Co Ltd, Scunthorpe, Lincolnshire.
(4) to Ind Coope & Allsopp Ltd, 6/1934 with works.

BAGGERIDGE BRICK & TILE CO LTD
BAGGERIDGE BRICK WORKS, Gospel End, Sedgley.   SJ1

Tramway laid from brick works (SO901932) to clay pits. Line closed 8/1957 and track removed by 3/1958.

Gauge: 2ft 0in

4wPM	MR	529	1917	(a)	(1)	
4wPM	MR	5324	1931	(b)	s/s /1957.	
4wDM	MR	5630	1932	(c)	s/s /1957.	
4wDM	MR	5853	1934	(d)	(2)	
4wDM	RH	186303	1937	(e)	(2)	

Another 4wPM was here, s/s after 4/1954, details unknown.

(a) ex WDLR 2250, orig delivered new for service in Egypt.
(b) ex J.Wilson, Tame Valley Canal Bridge reconstruction contract, Perry Barr, Birmingham.
(c) ex Tarmac Ltd, Gas and Electricity joint siding construction, Washwood Heath, Birmingham.
(d) ex Milton Hall (Southend) Brick Co Ltd, Thorpe Bay Brickworks, Essex after /1945.
(e) ex Charles Brand & Sons Ltd., contrs, previously Wansford Quarries Ltd, Northants.

(1) s/s after 4/1954.
(2) to Kingsbury Brick & Tile Works Ltd., Kingsbury, Warwickshire, c/1958.

BASS, MITCHELLS & BUTLERS LTD.
BURTON ON TRENT BREWERIES
Bass, Ratcliff & Gretton Ltd. until 1/4/1961
Bass, Ratcliff & Gretton & Co until 13/1/1888

HIGH ST BREWERY	SH23
MIDDLE BREWERY	SH2
DIXIE STORES	SH22
SHOBNALL MALTINGS	SH20
NEW BREWERY	SH29
MIDDLE YARD	SH28
ANDERSTAFF MALTINGS	SH40

Of all the breweries in Burton, those of Messrs Bass, Ratcliff and Gretton were to become the largest; the company's property ultimately extended over 800 acres around the town of Burton. Founded in 1777, Bass established a brewery in High Street (SK253231). In 1853 a second brewery was built in Guild Street (SK248233) nearer to the Midland Railway station. At the time much of the breweries traffic was taken through the streets to the station on carts locally known as floaters.

On the 4th October 1862, the Midland Railway opened the Guild Street branch for traffic and at the same time sidings were laid into Bass's Middle Brewery and the High Street brewery (Old Brewery). Until August 1865 much of the traffic on the Guild Street branch was handled by horses either owned by the Midland Railway or by one of the breweries served by the branch. Bass' horses worked traffic from Middle Yard and the Old Brewery until November 1863 when their first locomotive arrived. This engine then worked traffic along the Guild St branch to exchange sidings at Dixie beside the Midland Railway main line. In 1864 the New Brewery was completed (SK246229). Bass' private railway was extended across Station Street from the Middle Brewery into the New Brewery yard. Further extensions were made in the following year with the building of a line across Guild Street into the Middle Yard cooperage. Soon a large locomotive fleet was amassed to work the expanding system of private lines which served a nucleus of ale stores, breweries, cask washing plant and the Middle Yard cooperage. Running powers were also exercised over the extensive system of freight branches of the Midland Railway throughout the town, and continued into BR days, with certain locos being registered to do so.

The tide of expansion was continued between 1873 and 1876 with the building of ale stores and cask washing banks (SK248238) at Dixie. At Shobnall (SK234228) maltings, new bottling stores, ale stores and cask washing plants were constructed in 1873 - 1875. Shobnall sidings opened 9/11/1874 and traffic was worked by Midland Railway locos until April 1875, then Bass's locos took over. At first Bass engines ran from Dixie to Shobnall through the Midland Station. This practice was short lived, however. The Midland Railway Bond End branch was completed in 1875 and a connection was made with the New Brewery from 1/12/1875. Shobnall bound trains then ran direct along the Midland Railway Bond End and Shobnall branches. Maltings also existed in Anderstaff Lane (SK252237) and were reached from the Saunders Branch of the Midland Railway.

Some shunting was carried out by horses, also used for general cartage, throughout the railway's life. As late as 1926 approximately 120 horses were reported working at the Breweries. From 1928 Bass started to used tractors and in 1931 only 36 horses were at work. The locomotives were based at sheds in Guild Street (SK248234) The probable chronology of these is that No.1 shed was built in 1861 with an extension and fitting shop added in 1871. No. 2 Shed follwoed in the early 1870's and a third shed near to Scutari Maltings at the end of the 1890's. Two locomotives were also kept at Shobnall wharf (SK235230).

Bass instituted a form of classification for its locomotive fleet. The first five were unclassified but the rest were as follows:

Class Faery	0-4-0WT (5)	Thornewill and Warham
Class Triangle	0-4-0ST (2)	Thornewill and Warham
Class A	0-4-0ST (6)	Neilson Reid and NBL
Class B	0-4-0ST (2)	Rebuilt Class Faery (1) and Triangle (1)

Note also that a number of the locomotives carried plates of the form 'Rebuilt Bass, 1909'. These referred to normal overhauls which did not change the character of the locomotive, and so it is not appropriate to record them here.

The business of Worthington & Co Ltd was merged with Messrs Bass and stocks of Bass and Worthington locomotives were pooled and Worthington locomotives were numbered into the Bass series from 27/5/1960. The run down of the system started in early 1960 culminating in the closure of all Bass's loco sheds and the rail system throughout the New and Old Breweries. Beer was now pumped to a central loading point in the Middle brewery and the remaining sidings were shunted by BR. Bass's shed in Guild Street was closed in 1966 and rail traffic ceased entirely 5/1967. The last engines worked from the former Worthington shed. Withdrawn locomotives including some steam locomotives were also stored at the Worthington shed prior to sale.

Ref: A Notable Brewery Railway System, Bass, Ratcliff and Gretton, 1926.

Gauge: 4ft 8½in

No.1	0-4-0WT	OC	TW	224	1863	New	Scr.
No.2	0-4-0WT	OC	TW	249	1864	New	Scr.
No.3	0-4-0WT	OC	TW	259	1864	New	Scr.
No.4	0-4-0WT	OC	TW	303	1869	New	s/s c/1901
No.5	0-4-0WT	OC	TW	353	1872	New	(1)
	0-4-0WT	OC	TW	373	1873	New	
	Reb 0-4-0ST	OC			1897		(2)
No.6	0-4-0WT	OC	TW	393	1874	New	
	Reb 0-4-0ST	OC	HE		1900		(3)
No.7	0-4-0WT	OC	TW	400	1875	New	
	Reb 0-4-0ST	OC	HE		1899		
	Also Reb		HE		1915		(4)
No.8	0-4-0WT	OC	TW	420	1876	New	
	Reb 0-4-0ST	OC	HE		1898		(2)
No.9	0-4-0WT	OC	TW	425	1877	New	Scr.

No.2		0-4-0ST	OC	TW	455	1880	New	(5)
No.3		0-4-0ST	OC	TW	609	1890	New	
		Reb		Bass	1904, 1909, 1924			(6)
No.10		0-4-0ST	OC	NR	5567	1898	New	(6)
No.11		0-4-0ST	OC	NR	5568	1898	New	(7)
No.1		0-4-0ST	OC	NR	5759	1900	New	(4)
No.2		0-4-0ST	OC	NR	5760	1900	New	(7)
No.9		0-4-0ST	OC	NR	5907	1901	New	(8)
No.4		0-4-0ST	OC	NBH	19848	1913	New	(7)
No.12		0-4-0ST	OC	HC	452	1896	(a)	(9)
No.5		0-4-0DM		Bg	3027	1939	New	(10)
No.8		4wDM		RH	416566	1957	New	(11)
No.6		0-4-0DM		Bg	3509	1958	New	(12)
No.7		4wDH		S	10085	1961	New	(13)
No.12		4wDH		S	10003	1959	(b)	(14)
6		0-4-0ST	OC	HC	1417	1920	(c)	(15)
2		0-4-0ST	OC	HC	690	1904	(c)	(16)
13	(1 until /1961)	0-4-0ST	OC	WB	2815	1945	(c)	(6)
15	(4 until /1961)	0-4-0ST	OC	HC	602	1901	(c)	(4)
16	(5 until /1961)	0-4-0ST	OC	WB	2108	1923	(c)	(7)
17	(7 until /1961)	4wDM		KC		1924	(c)	(17)
18	(8 until /1961)	4wDM		KC		1924	(c)	(17)
19	(9 until /1961)	4wDM		KC		1925	(c)	(17)
20	(10 until /1961)	4wDM		KC		1926	(c)	(18)
21	(11 until /1961)	4wDM		FH	1612	1929	(c)	(18)
22	(12 until /1961)	4wDM		FH	1846	1934	(c)	(17)
		0-4-0DE		RH	412716	1957	(d)	(19)
No.1		0-4-0DM		Bg	3568	1961	New	(20)
No.4		0-4-0DM		Bg	3589	1962	New	(21)
No.11		0-4-0DM		Bg	3590	1962	New	(22)

(a) ex Worthington & Co Ltd., Burton, 6/4/1954.
(b) ex S, 5/1960. Previously used as a demonstration locomotive.
(c) ex Worthington & Co Ltd., Brewers, Burton, 27/5/1950.
(d) ex RH demonstration loco. 4/1959

(1) to ? , Sheffield, /1913
(2) to WD Purfleet, Essex, 20/4/1917.
(3) to S.Allsopp & Sons Ltd., Brewers, Burton, /1916.
(4) to Thos.W.Ward Ltd, Sheffield for scrap, 3/1963.
(5) to Sleaford Maltings, Lincolnshire, 10/1901.
(6) to Thos.W.Ward Ltd, Bolton on Dearne, Yorkshire (WR), for scrap, 5/1963.
(7) to Thos.W.Ward Ltd, Friargate Goods Yard, Derby for scrap, 8/1964; Scrapped 9/1964.
(8) to ? , Stafford, in store for Shugborough Hall Museum, Stafford, 17/2/1967
(9) scrapped on site by G.E.Baker (Metals) Ltd., Derby, 11/1958.
(10) to John Gretton, Melton Mowbray, Leicestershire, 8/1967. Transferred to GWR Preservation Society, Didcot, Oxfordshire, 31/8/1968.
(11) to L.Sanderson Ltd, dealers, Birtley, Co Durham, 3/1968. Resold to South Staffordshire Wagon Co Ltd, Tipton, /1969.
(12) to TH, Kilnhurst, Yorkshire, 11/1968, resold to Boulton & Paul Ltd., Lowestoft, Suffolk, 14/4/1969.

(13) to TH, Kilnhurst, Yorkshire 11/1968, resold to Marchon Products Ltd, Whitehaven, Cumberland, 8/1969.
(14) to TH, Kilnhurst, Yorkshire, 11/1968, rebuilt. Then dispatched to CEGB North Tees, Haverton Hill, Co Durham, 10/1969.
(15) to G.E.Baker (Metals) Ltd., Derby, for scrap, 10/1960.
(16) to G.E.Baker (Metals) Ltd., Derby, for scrap, 6/1961.
(17) to Albert Looms Ltd, Spondon, Derbyshire, for scrap, 8/1967.
(18) to Railway Preservation Society, Chasewater, Staffs, 7/1967.
(19) returned to RH /1959.
(20) to Corralls Ltd, Dibbles Wharf, Northam, Hants, 2/1968.
(21) to Geo Cohen, Sons & Co Ltd,, Canning Town, Essex, 9/1967.
(22) to Wagon Repairs Ltd, Port Tennant, Glamorgan, 3/1968.

One TW locomotive was also sold to Hemmingways Chilled Rolls Ltd, Haverton Hill, Co Durham and subsequently sold c/1906.

BIRMINGHAM CANAL NAVIGATION
LITTLEWORTH TRAMWAY
SC1

Though the Birmingham Canal Navigation had several narrow gauge tramways, only one standard gauge line was built for them. The Littleworth tramway construction commenced about 1863 and was carried out for the LNWR presumably by the same contractors who built the Cannock Chase Railway. The line was built from an end-on junction with the Cannock Chase Railway (Wimblebury Junction) to the wharves and basins at the end of the Cannock Extension Canal. When complete the BCN paid the LNWR for the construction costs and commenced to charge a toll for goods carried over it. At first traffic along the line was to Mr Piggott's Hednesford colliery, and LNWR engines would work over BCN metals to reach the mine. Later agreements were made with the Cannock & Rugeley Colliery Company and the Cannock Chase Colliery Co Ltd to work their locomotives over the line. The CRC worked traffic to Hednesford basin and the CCWR working to the Hednesford Colliery. About 1883 an agreement was also made with the New Cannock & Wimblebury Colliery Co for working traffic from its Wimblebury colliery to the LNWR over the Littleworth Extension Railway (which see). The traffic on this line was mainly handled by the CCWR and the CRC and each was restricted to working the line at different times of the day. These workings remained in force well into NCB days. Coal traffic ceased about 1963, a hundred years after the line was made. A system of de-mountable containers was used for house coal from Cannock Wood screens. Loaded three to a flat wagon, they were transferred by crane at the wharf. This system remained in use until 1961, and probably until closure.

BRANSTON ARTIFICIAL SILK CO LTD
BRANSTON WORKS, Burton on Trent.   SH4

This company was established 1927 and used the former Crosse & Blackwell Branston factory (SK233215), originally MOM Branston. Sidings connected with the LMS Burton to Birmingham line south of Burton. Factory closed 1930 and was requisitioned by the War Department in 1937. Later these premises became ROF Branston, (which see).

Gauge: 4ft 8½in

No.1   0-4-0PM   BgE   1654   1928 New   (1)

(1)   to WD Branston with premises 7/1937.

BRANSTON GRAVELS LTD
BRANSTON GRAVEL PITS.   SH16
H & E.W Harrison until /1931.

These gravel pits (SK215203) used locomotive haulage until the railway was replaced in 1958 by a conveyor belt system.

Gauge: 2ft 0in

4wDM	RH	166022	1933	New	(1)
4wDM	RH	174548	1935	New	(2)
4wPM	MR	5206	1931	(a)	s/s.
4wPM	MR	5401	1931	(b)	s/s.

(a)   ex Queslett Sand & Gravel Co Ltd, Great Barr, Staffs.
(b)   ex Midland Gravel Co Ltd., Water Orton Pits, Warwickshire.

(1)   to Midland Gravel Co Ltd., Booths Farm Pits, Great Barr, Staffs.
(2)   to Midland Gravel Co Ltd., Kingsbury, Warwickshire.

BRERETON COLLIERIES LTD
BRERETON COLLIERIES, near Rugeley   SA1
Earl of Shrewsbury's Brereton Collieries Ltd. until 9/1920
Earl of Shrewsbury and Talbot, Brereton Collieries. until 9/1906
Lord Talbot until c/1861

Brereton Colliery (SK048159) was first served by a tramroad c1200 yds long laid in 1815 and worked by horses to a basin (SA3) beside the Trent and Mersey Canal at Lea Hall near Rugeley. In 1854 The Hayes Colliery (SK041151), whose workings bordered on those of Brereton, was purchased from the Marquis of Anglesey. Hayes Colliery had its own tramroad, which ran through the streets of Rugeley, to the Trent and Mersey canal ; but this tramroad was removed sometime shortly after the sale. In 1859 the Cannock Mineral Railway was built from Cannock to Rugeley and a branch was started from the Brereton Collieries to the new line the same year. Little work was done, however, and it was not until 1875 that a more serious attempt was made. This time the colliery railway **(SA2)** was completed. Six pits were worked by the company; Belfast (SK048161) **(SA10)**, Brereton, Brick Kiln (SK044152) **(SA11)**, Coppice (SK047163) **(SA12)**, Hayes **(SA13)** and Old Engine (SK049155) **(SA14)** and all were served by the line which climbed steeply to exchange sidings beside the LNWR Cannock Mineral Railway south of Rugeley Town. Locomotive sheds were erected between Hayes and Brereton pits (SK045154). There were also brickworks set up beside the line and one near The Springs was served by a narrow gauge tramroad that ran 400 yards west to a wharf beside the colliery railway at Old Engine Pit. During the 1880's the horse worked tramroad to the Trent and Mersey Canal was replaced by a rope worked line with the haulage engine placed at The Levels near to the Brereton mine. This line remained in use until 1922. Locomotives were, with one exception, four coupled, but nevertheless had to be powerful for the steep ascent to the exchange sidings, and usually two locomotives were employed on the line at any one time.

The colliery at SK 044152 **(SA11)** passed to the NCB (which see) as Brereton Colliery on 1/1/1947.

Ref: A Transport History of Cannock Chase- R.Francis.

Gauge: 4ft 8½in

1		0-4-0ST	OC	HCR	161	1875	New	s/s.
2		0-4-0ST	OC	HCR	194	1878	New	s/s.
3		0-4-0ST	OC	HCR			(a)	s/s.
(No.2)	BARBARA	0-4-0ST	OC	AB	1083	1908	New	Scr/1935
(No.3)	4	0-4-0ST	OC	AB	1115	1909	New	Scr/1943
	VANGUARD	0-4-0ST	OC	P	1491	1917	(b)	(1)
No.3		0-6-0ST	OC	MW	1852	1914		
		Reb	RHLM	1924			(c)	Scr/1935
No.3		0-4-0ST	OC	AB	1365	1914	(d)	(1)
No.8		0-4-0ST	OC	Heath		1888		
		Reb /1891 and CW /1931					(e)	(2)
B2C		0-4-0ST	OC	Butt		1889	(f)	(1)

(a) identity unknown. Believed purchased second hand.
(b) ex Royal Arsenal, Woolwich, /1920.
(c) ex Norton & Biddulph Colls Ltd, Victoria Colliery, N. Staffordshire, /1930.
(d) ex Naworth Coal Co Ltd., Cumberland, /1935.
(e) ex Norton & Biddulph Colls Ltd, loan, /1942.
(f) ex Butterley Co Ltd., Codnor Park Forge, Derbyshire.

(1) to NCB, WM Division, Area 2, 1/1/1947, with colliery.
(2) retn to Norton & Biddulph Colls Ltd.

BRITISH WATERWAYS
NORTON CANES CANAL WHARF.   SE6
Birmingham Canal Navigation Co Ltd until 1/1/1948.

The locomotives were used at the canal wharf (SK020079) and on canal maintenance where required. Track removed by 3/1960.

Gauge: 2ft 0in

	4wPM	L	962 c/1930	New	(1)
P109	4wPM	L	26060 1944	New	Scr c/1960

(1) later at E.Evans (Steel and Metal) Ltd, St Vincent Street, Birmingham, then resold to A.Keef, Cote Farm, Bampton, Oxon, /1983. It is not known how long this loco remained in the possession of BCN, whether it went direct from BCN to E Evans, or whether there were other owners in the interim.

BURTON BREWERY CO LTD
BURTON ON TRENT BREWERY. SH5
Henry and Thomas Wilders until /1858

The Midland Railway opened its Hay Branch as far as Anderstaff Lane in November 1861 and then gradually increased the length of this branch until it reached The Hay on November 6th 1864. Until this date no locomotive crossed Anderstaff Lane and Midland Railway horses hauled the traffic to and from the breweries served by them at this part of the town. The Burton Brewery Co had its brewery (SK254233) in High Street. It had been founded in 1842 and by 1861 was the third largest brewery in Burton. Rail traffic commenced on 3/2/1863 and construction of a siding to the hay Branch was authorised in 8/1867. The company purchased a locomotive in 10/1867 to deliver its traffic into the hands of the Midland Railway. This engine and subsequent locomotives owned by Burton Brewery worked along the Hay and Saunders Branches to maltings and the company's cooperage (SK252239) **(SH25)** which stood between Hawkins Lane and Anderstaff Lane. Premises acquired by Worthington & Co Ltd, 1914.

Gauge: 4ft 8½in

No.1	0-4-0ST	OC	MW	228	1867	New	s/s.
No.2	0-4-0ST	OC			1872	New	s/s.
No.3	0-4-0ST	OC	MW	593	1877	New	(1)
No.4	0-4-0ST	OC	MW	1427	1899	New	(2)

(1) to J.D.Nowell and Sons, contractors, Warrington, Lancs.
(2) to Worthington & Co Ltd., Brewers, Burton on Trent, /1914.

BURTON CONSTRUCTIONAL ENGINEERING CO LTD
DERBY ROAD WORKS, BURTON ON TRENT SH6
(Company registered 5/1914. Subsidiary of Marple & Gillott Ltd by 1924)

These works (SK256248) were established in 1914. An extensive narrow gauge railway was laid within the confines of the works serving the workshops and stockyard. Rolling stock was apparently home made and comprised four wheeled flat chasis wagons. Rail system almost entirely removed after 1972 and at the same time locomotive haulage was dispensed with.

Gauge: 2ft 0in

	4wPM	FH ?			(a)	s/s.
	4wDM	FH	1869	1934	(b)	s/s c/1960.
	4wDM	HE	2176	1940	(c)	(1)

(a) identity uncertain; may be either a MR or FH.
(b) ex First National Housing Trust Ltd., Perry Beeches Estate construction, Birmingham.
(c) ex Geo Cohen Sons & Co Ltd, Kingsbury, Warwicks, c/1960; orig WD.

(1) to Leighton Buzzard Narrow Gauge Railway Society, Bedfordshire, property of J. Thomas, 9/1972.

CANNOCK CHASE COLLIERY CO LTD
CANNOCK CHASE COLLIERIES, Chasetown. SC6-10
Cannock Chase Colliery Co until 28/05/1902. SE1-5
Messrs McClean and Chawner until /1859.
Marquis of Anglesey until 4/1854.

The Cannock Chase Colliery Co assigned numbers to its pits instead of naming them, as was the common practice in the Midlands, yet some still had local nicknames all the same. Ultimately, there were ten pits numbered in the Cannock Chase sequence and these were either sunk or purchased between 1849 and 1870. All ten were never in production at the same time and some had shorter lives than others. From 1870 no new pits were added to the group and production concentrated on the remaining six pits.

The No.1 Pit (SK041074) **(SE1)**, originally known as Hammerwich Colliery, was sunk in 1849 for the Marquis of Anglesey beside the canal feeder from Chasewater. This piece of water was made a navigable canal (Anglesey Canal) in 1850 to serve the new mine. In 1852 the No.2 Pit (SK039082) **(SE2)** , also known as Uxbridge Pit, was sunk to the north of the Hammerwich mine and a standard gauge mineral railway was laid to link the Uxbridge and Hammerwich Pits with the South Staffordshire Railway Co at a spot later to be known as Anglesey Sidings. A locomotive called BLACKCOCK was purchased to haul pit tubs in specially adapted wagons from Uxbridge to a wharf beside the Anglesey Canal and conventional wagons to the SSR.

In 1854, these two mines were leased to John McClean (who was the Lessee and Engineer for the SSR at the time) and Richard Chawner (a SSR director). The partners traded as the Cannock Chase Colliery Company. Within the next three years another three pits (3-5) **(SE3-5)** were sunk to the north and north east of the Uxbridge Pit and linked to the SSR through an extension of the existing mineral lines.

John Mc Clean gave up his lease of the SSR to the LNWR in 1861 and soon was interested in other railway projects. He was chiefly responsible for a new railway scheme to connect his collieries to the GWR at Wolverhampton. Called the Cannock Chase & Wolverhampton Railway (CCWR), powers were granted for its construction in 1864. The line's course left the existing railway between No.3 (SK035092) and No.4 (SK041084) collieries, skirted Chasewater and crossed the LNWR Norton Branch, where a connection was to be made. It then continued through Norton Canes, Essington and Wednesfield to join the GWR at Cannock Road Junction. A second Act was passed in 1866 which absorbed the railway from Anglesey Sidings into the CCWR and authorised a new railway from No.3 Colliery **(SE3)** through Chase Terrace to join the Cannock Chase Railway at Cooper's Lodge, near Rawnsley.

Land was compulsorily purchased and construction commenced on the railway to the Norton Branch and to Rawnsley, between 1866 and 1868, but no work was undertaken on the main railway from Norton to Wolverhampton, which was abandoned in 1869. The CCWR remained in business to operate the railways which served the Cannock Chase Colliery Company mines, a lasting reminder of hopes that were never realised.

Meanwhile further mines had been sunk. No.6 (SK 018111) **(SC6)** was a short lived cannel pit near Wimblebury which opened in 1866 and was linked to the main system by a branch railway from No.3 Pit. This line was not included in the 1866 Act and would have remained under the ownership of the Cannock Chase Collieries.

No.7 Pit (SK037111) **(SC7)** was completed in 1868 and was sited alongside the CCWR extension from Chase Terrace to Rawnsley.

The No.8 Colliery (SK021106) **(SC8)** was the last new sinking to be made and it was to become one of the largest in the group. Situated at Wimblebury, this mine stood near the end of the branch which was built to the No.6 mine.

No 9 (SK009112) **(SC9)** and No.10 (SK005113) **(SC10)**, purchased in 1870, formed what was known as the Hednesford Colliery. This mine was first worked by Thomas Piggott who traded as the Hednesford Colliery Co. Mr Piggott commenced operations by leasing the property in August 1858 from Lieut. Colonel Lovett. Sidings from the mine connected with the BCN Littleworth tramway by which coal could either travel onto the BCN coal wharves at Hednesford to be distributed by canal or via the Cannock Chase Railway onto the LNWR at Hednesford. From 1881, the spur known as the Littleworth Extension Railway made a connection with the LNWR Norton Branch. Though strictly a LNWR line, the Littleworth Extension Railway (which see) was worked by the Cannock Chase Colliery Co.

No 1 Pit was the first to close (probably by 1860) followed by No.6, (1874) and then No.4 (by 1880).

In 1882 the Midland Railway completed its mineral railway from Brownhills to Chasewater and a new connection was made with the CCWR. Later the junction with the LNWR Norton Branch was removed when traffic for this route was forwarded over a new mineral line to the Five Ways branch at Conduit Colliery.

Locomotives were originally based at No.2 Pit and worked over the whole system including the Hednesford Colliery mines. At No.2 there were also workshops, a forge and a wagon works. There were two brickworks; one was situated at the No.7, the other was at No.10, although there were small brickmaking units at most mines in the group.

In 1924 considerable modifications were made to the plant at the mines. No.2 Pit ceased coal winding and the method of hauling tubs on bogies from pit mouth to canal wharf was replaced by an electrically driven endless rope haulage system connecting a new drift to the underground workings with the wharf, a distance of 960 yds. A policy of centralisation was also introduced in 1924 concentrating activities at the No.3 Pit site **(SE3)**. Central workshops, locally known as the 'Wembley' were opened in that year on land adjacent to the No.3 mine and replaced the old workshop and foundry at the No.2 Pit **(SE2)**. Running sheds for the locomotives remained at the site of the No.2 Pit.

Mining at the smaller pits was also discontinued. No.10 **(SC10)** closed by 1910 and No.5 (SK042093) **(SD5)** had closed by 1922. The survivng plant became increasingly interconnected. In 1927 an underground drift (1½ miles in length) was constructed linking No.8 with No.3 Pit and later in the same year a new surface rope haulage line (gauge 2ft) was built to link these two pits. The intention was to bring coal in mine tubs from the No.8 Mine to blend with other coals at the extensive screening plant at the No.3 Mine and thereby avoid the time consuming task of transhipping coal from standard gauge wagons. Between 1927 and 1928 other underground drifts were driven; No.9 was connected to No.8 and No.3 linked with No.2. For a number of years, between 1928 and 1935, No.9 ceased coal winding and all the coal was taken underground to No.8. The brick making was confined to the plant at No.10 (**SC10**).

No.3, No.8 and No.9 Collieries together with almost all the rail system survived to become vested in the NCB. Many of the locomotives dated back to the early years of the undertaking and it is to the credit of the workmanship of the company's engineers that most were still running at Nationalisation.

Underground haulage was by horses, endless ropes and main and tail systems with power being provided by steam and later electricity.
The surface railways were equipped with semaphore signals.

Ref: The Mining Journal, p565 1853.
      The Colliery Guardian, 12/12/1924.
      The Locomotive, - Oct 14th 1939, pp297/298.
      The History of Cannock Chase Colliery Company- Rodger Francis.

Gauge: 4ft 8½in

	BLACKCOCK	0-4-0WT					(a)	s/s.
	McCLEAN	0-4-2ST	IC	BP	28	1856	New	(1)
	ALFRED PAGET	0-4-2ST	IC	BP	204	1861	New	(1)
	CHAWNER	0-4-2ST	IC	BP	462	1864	New	(1)
	BROWN	0-4-2ST	IC	BP	794	1867	New	(2)
	ANGLESEY	0-4-2ST	IC	BP	1211	1872	New	(1)
No.6		0-6-0ST	IC	SS	2643	1876	New	(1)
	GRIFFIN	0-6-0ST	IC	K	5036	1913	New	(1)
	FOGGO	0-4-2ST	IC	Chasetown		1946	New	(1)
75070		0-6-0ST	IC	RSH	7106	1943	(b)	(1)

(a) identity and origin unknown, presumed here new.
(b) ex WD 75070, 12/1946.

(1) to NCB, WM Division, Area 2, with mines, 1/1/1947.
(2) Substantially dismantled in 1926 and some parts utilised to maintain the existing fleet.The rest was scrapped.

No.6 worked with a four wheeled wagon as a tender and was used on the longer haul duties such as to No.9 and 10 Pits. FOGGO was built at Chasetown workshops using parts supplied by BP 3/1946 and spare parts accumulated over the years for the other locomotives. The name is that of the General Manager at the time - M.J.Foggo.

THE CANNOCK CHASE RAILWAY SC2

To present a complete picture of the Cannock Chase industrial scene, mention must be made of the above line. Powers for its construction were granted to the Marquis of Anglesey in 1860. The line was projected to run from Hednesford to Coopers Junction (LNWR/CCWR) and then a switch back took the line onto Heathy Leasowes to make an end on junction with the BCN Littleworth tramway. Built by Brassey and Field for the LNWR after 1863, it was purely a mineral line serving brickworks and collieries. Locomotives of the Cannock & Rugeley Co Ltd worked along its entire length and the Cannock Chase Colliery locomotives also worked along part of its route to reach the Hednesford mines. When both colliery companies were vested in the NCB the practice continued right up to the closure of the last working colliery, Cannock Wood, in 1973. The line was then closed and lifted.

CANNOCK CHASE & WOLVERHAMPTON RAILWAY CO LTD

Incorporated by Act of Parliament 29/07/1864 which authorised a railway from the Cannock Chase No.3 Mine through Essington and Wednesfield to connect with the GWR at Cannock Road Junction, Wolverhampton. A second Act dated 16/07/1866 gave powers to purchase the existing colliery railway from Cannock Chase No.3 Mine to Anglesey Sidings where a junction was made with the LNWR and also to build a new railway from Cannock Chase No.3 Mine to join with the Cannock Chase railway at Cooper's Junction.

These railways were authorised mainly through the efforts of John McLean, whose intention was to provide a through route for Midland Railway traffic to Wolverhampton as well as providing a further outlet for his coals. The Midland Railway had running powers over the South Staffordshire Railway and the new line would have provided it with a direct route. However the Midland Railway obtained authority to work over the Wolverhampton & Walsall Railway which weakened its support for the construction of the new line.

Most of the railway authorised by the 1866 Act was completed by 1868. Captain Tyler's report to the Board of Trade in 1869 stated that the lines had handled only coal traffic and had been worked exclusively by engines, wagons and vans belonging to the colliery proprietors. On the 30th November 1869 the Board of Trade granted an abandonment order for the line from Norton to Wolverhampton.

The CCWR remained in existence as a separate company even though its operations were closely involved with the Cannock Chase Colliery Company (which see). The CCWR maintained the track and provided a semaphore signalling system to aid operations, and also maintained and operated the locomotives. In 1947, when the Cannock Chase Colliery Co was vested in the National Coal Board, the CCWR was also acquired, but as a separate concern. On 1st Jan 1947 the NCB took over 5 miles 794 yards of single track railway, 2 miles 942 yards of sidings, 9 signal boxes, 8 level crossings and 12 signal posts. They also acquired the locomotive sheds at Chasetown, 8 locomotives and numerous spares.

See under the Cannock Chase Colliery Co Ltd for details of the locomotives.

CANNOCK & RUGELEY COLLIERY CO LTD
CANNOCK WOOD COLLIERY	SC3
VALLEY or POOL COLLIERY	SC4
WIMBLEBURY COLLIERY	SC5
RAWNSLEY DEPOT	SC19

This company was promoted and registered 19/5/1864 (re-registered 14/2/1865) to sink shafts and open up a new part of the coalfield under Cannock Chase. Sinking was commenced in the mid 1860's on land leased from the Marquis of Anglesey and coal measures were finally reached in June 1866. Two shafts were sunk on the site which became known as the Cannock Wood Colliery (SK033126). These shafts were 12ft in diameter and had been sunk to their full depth of 200 yards by August 1866. A third, of 16ft diameter, was later sunk on the same site. A mineral line was laid by 1866 to link the mines with the Cannock Chase Railway, which in turn joined the Cannock Mineral line at Hednesford.

At first the CRC hired locomotives from I.W. Boulton of Ashton under Lyne to deliver their coal traffic to the LNWR, but from 1867 commenced to use its own locomotives. These worked either over the Cannock Chase Railway to Hednesford or by the Littleworth Extension Railway to reach the wharves of the Cannock Extension Canal.

In 1873 a new sinking was started at the Valley or Pool colliery (SK008127) where production commenced in 1875. When completed this pit had shafts 15 feet in diameter sunk to a depth of 350 yds. A short branch railway linked these pits with the LNWR sidings at Hednesford which CRC locomotives reached by running over the Cannock Chase Railway.

The CRC built its locomotive sheds beside the Cannock Chase Railway at Rawnsley (SK025124) and, workshops were built in-between the junction of the two arms of the Cannock Chase Railway at Rawnsley. The locomotive BIRCH was built at these workshops in 1888. The next year (1889), the CRC took possession of the Wimblebury Colliery (SK014117) which stood beside the Littleworth branch. The colliery had been idle since 1887 when the previous company to work it, the New Cannock & Wimblebury Colliery Co Ltd., had gone into liquidation. Wimblebury and Valley were worked as one unit by CRC from 1889 with all coal being wound at Wimblebury while Valley was used for man-riding and materials. Rail access to Valley was retained until NCB days.

A passenger service for the benefit of CRC workers was provided which ran from Hednesford to Cannock Wood Colliery where a special platform was built. This practice continued into the 1960's. The stock was unusual and latterly antiquated: one was a Furness Railway six wheeled coach, the second a Maryport and Carlisle six wheeled coach while the third was a GER brakevan. The CRC also had a large fleet of coal wagons which by 1900 numbered 1450. All were repaired at Rawnsley and some were constructed there.

Ref: Engineer, 6/6/1873.
        The Mining Journal, 27/1/1883.
        Colliery Guardian, 5/5/1893.
        The Locomotive, 15/2/1940
        CRC Ltd. 12/1946 (Historical Summary in British Coal archives)
        The Journal of the Stephenson Locomotive Society, 11/1953

Hired locomotives (possibly used on construction):
Gauge: 4ft 8½in

	OPHIR		2-2-0ST	3C	GE		
		reb to	0-4-0ST	IC	I.W.Boulton	(a)	(1)
26			0-4-0	IC	Bury		
		reb to	0-4-0ST	IC	I.W.Boulton	(b)	(2)

(a)   ex I.W.Boulton, hire, 30/12/1866.
(b)   ex I.W.Boulton, hire, by 4/1867.

(1)   retn to I.W.Boulton by 4/1867. Sold to E.Knight, contrs, South Milford, Yorks (WR).
(2)   retn to I.W.Boulton. Later sold to Jamieson and McCormack, Wigan.

Permanent Stock:
Gauge: 4ft 8½in

	1	MARQUIS	0-6-0ST	IC	Lill		1867	New	(1)
	2	ANGLESEY	0-6-0ST	IC	Lill		1868	New	(1)
	3	UXBRIDGE	0-4-0ST	OC	Lill		1868	New	(2)
		CANNOCK WOOD	0-6-0ST	IC	Lill		1870	New	(3)
	4	RAWNSLEY	0-6-0ST	IC	Lill		1872	(a)	(1)
	5	BEAUDESERT	0-6-0ST	OC	FW	266	1875	New	(1)
	6	CANNOCK WOOD	0-6-0ST	OC	FW	318	1876	New	(4)
	7	BIRCH	2-4-0T	OC	Rawnsley		1888	(g)	(1)
	8	HARRISON	2-4-0T	OC	YE	185	1872		
		reb to	0-6-0T	OC	Rawnsley		1916	(b)	(1)
(3)		MESSENGER	0-6-0ST	IC	MW	166	1865	(c)	(5)
	3	PROGRESS	0-6-0ST	IC	P	786	1899	(d)	(1)
	6	ADJUTANT	0-6-0ST	OC	MW	1913	1917	(e)	(6)
	9	CANNOCK WOOD	0-6-0T	IC	Bton		1877	(f)	(1)

(a)   ex Lilleshall Iron Co, Shropshire, 7/1873, rebuilt from a 2-2-2 built 1867 for the Paris Exhibition.
(b)   ex B.P.Blockley, contractor and dealer, Bloxwich, /1905. Previously No.1 of the East and West Junction Railway.
(c)   ex Braddock and Matthews contractors, c/1900.
(d)   ex Swansea Harbour Trust, 6A, c2/1915.
(e)   ex Admiralty, Beachley Dock, Gloucestershire, /1924.
(f)   ex Southern Railway, B110, 5/4/1927.
(g)   Built new in 1888 but not put to work until 1890

(1)   to NCB, WM Division, Area 2, with mines, 1/1/1947.
(2)   sold 7/1892.
(3)   to Walsall Wood Colliery Co Ltd, /1882.
(4)   to Holditch Mines Ltd, Chesterton, Staffs, /1927.
(5)   to West Cannock Colliery Co Ltd, c/1914.
(6)   to Littleton Collieries Ltd, loan, 9/1945.

Underground haulage:

The company was quite progressive, adopting modern techniques and the latest equipment. Power was provided for haulage engines by steam boilers, compressed air and later electricity.

Cannock Wood Colliery SC3
Underground haulage was by endless rope, main and tail and one section used a locomotive.

Gauge: 2ft 6in

                      0-4-0CA   Grange   1883   New   s/s.

Valley and Wimblebury Collieries
Underground haulage at both mines was by endless rope.

CANNOCK & WIMBLEBURY COLLIERY CO LTD
Incorporated 12/7/1873
WIMBLEBURY COLLIERY SC5

The Cannock and Wimblebury sinking (SK014117) struck coal in July 1874 and two shafts 155 yds deep and 11ft in diameter were complete by the end of that year.. The mine was located alongside the Cannock Chase Railway Littleworth branch to which sidings were soon made. Coal from the mine was conveyed either over the Littleworth tramway to the BCN Cannock Extension canal or over the Cannock Chase Railway to Hednesford for distribution via the LNWR. Pit horses were used underground for haulage. The pit gauge was 2ft 2in. The company went into voluntary liquidation 18/11/1880 and the mine was offered for sale in 1881 with the land, colliery fixed plant and brickworks being auctioned on the 21st April. The remainder came up for sale on the 25th and 26th of April and comprised the loose plant including pit horses, 200 pit tubs, 94 railway wagons and one Black Hawthorn locomotive. A new company called the New Cannock & Wimblebury Colliery Co Ltd. (which see), then took over the working of the pit .

Gauge: 4ft 8½in

    NORTH   0-4-0ST   OC   BH   363   1875 New   s/s.

CHARRINGTON & CO LTD
ABBEY BREWERY, Burton on Trent. SH7
Charrington & Co until 15/7/1897

Messrs Charrington constructed a new brewery on the site formerly occupied by Meakin's London and Burton Brewery Co in Abbey Street, Burton (SK248225). After the building of the Midland Railway's Bond End branch in April 1880, several extensions were made into the yards of breweries near to its route. One such line became known as the Robinson Brewery branch- more usually known as New Street Branch- and had a siding which crossed Lichfield Street into the Abbey Brewery. Messrs Charrington commenced to work their traffic with their own engine 1/5/1885 along the Midland Railway Bond End and Shobnall branches and traffic for the LNWR was taken as far as Dallow Lane Sidings. The maltings were located at Wood St (SH27). A short railway connected these with both of the Midland Railway's Bond End and Robinson Brewery branches. The locomotives were housed in a small one road shed (SK 243223) within the maltings and beside Queen Street. Brewery closed.

Gauge: 4ft 8½in

0-4-0ST	OC	HC	276	1885	New	(1)
0-4-0ST	OC	HC	1437	1921	New	(2)

(1) to HC, 3/1921; then resold by them to Taf Fechan Water Supply Board, Pontsticill Reservoir Contract, Glam., by 9/1927.
(2) to Joseph Crosfield & Sons Ltd., Warrington, Lancashire, /1926.

CONDUIT COLLIERY CO LTD
CONDUIT Nos 1, 2 and 3 PITS   SE30
JEROME (No.3) COLLIERY   SE7
NORTON GREEN (No.4) COLLIERY   SE8
Conduit Colliery Co until 26/06/1928

The various pits which collectively bore the name of Conduit Colliery were to be found at Brownhills and Norton Canes principally on lands leased from William Hanbury. Several of these mines were already in production by 1860.

Conduit No 1,2 and 3 Pits (SK023067) were sunk near to Watling Street at Brownhills. A tramroad 400yds long linked these mines with a basin on the east side of the BCN Cannock Extension Canal. Laid around 1863 this tramway would have been horse worked, but a study of the ordnance survey reveals that by the early 1880's a form of rope or chain haulage was in use. By 1900 all three of these pits had closed and the track had been lifted. Two larger pits were sunk in the neighbourhood of Norton Canes known as Jerome (No.3) and Norton Green (No.4) collieries.

Both were served by standard gauge sidings which linked with the SSR/LNWR Norton branch. Norton Green Colliery (SK016075) stood at the original terminus of the Norton branch and was closed temporarily from sometime in the 1890's, reopening again about 1910. Jerome was placed close to Norton village beside the Hednesford Road. The LNWR constructed extensive sidings next to this pit and the BCN also made a large canal basin adjacent to the mine. Later a mineral line was laid from the sidings opposite Jerome to join the Midland Railway near Brownhills.

Locomotives were based at Jerome Pit (SK020082) where a short one road shed and a long one road shed were located. Locomotives also shunted Norton Green returning to Jerome in the evening. Running powers were exercised over LNWR/LMS Five Ways Branch between Conduit Colliery Sidings, New Conduit Sidings and Conduit Junction and also along the Norton Branch to Norton Green. In 1931, the mines were taken over by Littleton Collieries Ltd.,(which see). The Conduit Colliery Co Ltd was wound up in 1932.

Gauge: 4ft 8½in

CONDUIT No.1	0-6-0ST	IC	MW	244	1867	New	(1)
CONDUIT No.2	0-6-0ST	IC	MW	565	1876	New	(2)
CONDUIT No.3	0-6-0ST	IC	MW	1180	1890		
		Reb	MW		1920	New	(1)
CONDUIT No.4	0-6-0ST	IC	MW	1326	1896	New	(1)

(1) to Littleton Collieries Ltd. with mines, /1931.
(2) to MW, c11/1891. Reb /1892. Resold to Ackton Hall Colliery Co, Yorkshire (WR) 3/1892.

South Staffordshire Handbook. Page 54

COPPICE COLLIERY CO LTD.
Coppice Colliery Company
J.Owen & Co until /1881
R.O.Hanbury (J.Owen, Chartermaster)

NO1- 5 PITS, BROWNHILLS COLLIERIES.   SE9

Working of coal on Brownhills Common commenced at an early date and by 1834, a tramroad was in place which started at mines near to Watling Street, crossed Coppice Lane and ran onto a wharf beside the Wyrley and Essington Canal at the Slough (SK034055). Later, with the building of the SSR Norton Branch a standard gauge siding was laid to the pits. Another canal tramroad was also built which ran due south from the No.1 (SK036063) and No.5 Pits for about 1000yards to a new basin beside the Wyrley and Essington canal. A locomotive called TINY is known to have worked upon this latter line. By 1900 all track had been removed and the locomotive(s) disposed of.

Gauge: Narrow

TINY	0-4-0T	(a)	s/s.
(ALICE) ENTERPRISE	0-4-0T	(a)	s/s.

(a)   probably acquired second hand. The names are similar to those carried by locomotives formerly employed by the New British Iron Co.

COPPICE NO.6 AND NO.8 PITS, Wilkin, Brownhills.   SE10

A more substantial plant was laid down at these two mines which were to be found north of Watling Street beside Wilkin village (SK031068). Standard gauge sidings were laid to meet the LNWR Norton Branch and were worked by a locomotive. These pits were disused before 1900.

Gauge: 4ft 8½in

(HANBURY) FAIR LADY	0-4-2ST	IC	BP	1915	1879	New	(1)

(1)   to Coppice Colliery, Heath Hayes after /1893.

COPPICE COLLIERY, Heath Hayes   SE11

Sinking of this new mine (SK014096) was started on Mr R.W. Hanbury's land in 1892 and was completed in 1893. It was also known locally as "The Fair Lady Pit", after Mrs Ellen Hanbury, wife of owner R W Hanbury. In 1894 the LNWR built a short branch (Five Ways Branch) from Conduit Junction to the colliery. Locomotives owned by the colliery shunted the mine and had running powers to work along the Five Ways branch to Conduit Colliery. They also worked over the mineral branch to the Midland Railway beside Chasewater. A 2ft gauge endless rope worked, double track tramway about 1 mile long was laid to take pit tubs from the mine to a wharf beside a canal basin (SK012083) on the Cannock Extension canal at Norton Canes. Screens were built alongside the canal basin. Underground haulage in the mine was also by endless rope. On 1/1/1947 the mine passed to NCB WM Div Area 2.

Gauge: 4ft 8½in

	HANBURY	0-6-0ST	IC	P	567	1894	New	(1)
	FAIR LADY	0-4-2ST	IC	BP	1915	1879	(a)	Scr /1926.
(2)	THOMAS	0-6-0T	IC	K	5358	1921	(b)	(1)

(a) ex Coppice Colliery, Wilkin, after /1893.
(b) ex Glyncorrwg Colliery Co Ltd, Glamorgan, No 2, 2/1926.

(1) to NCB, WM Division, Area 2, 1/1/1947, with mine.

CRODA SYNTHETIC CHEMICALS LTD
FOUR ASHES WORKS. SB1
Croda Hydrocarbons Ltd. until
Midland - Yorkshire Tar Distillers Ltd until /1976
Midland Tar Distillers Ltd until 12/1967

Works were erected in 1952 beside the BR Wolverhampton to Stafford line. Sidings connect the plant (SJ917085) with BR north of the site of Four Ashes station. The locomotive was occasionally used until 1991.

Gauge: 4ft 8½in

(GAMMA) MP1   0-4-0F   OC   AB   1944   1927   (a)   (1)

(a) ex Reckitt & Colman Ltd, Norwich, via Geo Cohen, Sons & Co Ltd, c11/1952.

(1) to Telford Horsehay Steam Trust, Shropshire, c7/1992.

CROSSE & BLACKWELL LTD
BRANSTON WORKS, Burton on Trent. SH4

The factory site was developed during the first world war as MoM Branston (which see). After the war it was purchased by Messrs Crosse & Blackwell for conversion into a food factory. However restrictions placed upon them by the MoM, meant that part of the plant had to be retained in case of another war. Unwilling to be bound by this constraint, Messrs Crosse & Blackwell chose to sell the premises, which were advertised for sale in February 1925. They were then purchased by the Branston Artificial Silk Company, (which see).

Gauge: 4ft 8½in

DAISY	0-4-0ST	OC	HCR	139	1874		
		Reb	HC		1914	(a)	(1)

(a) ex H.Symington & Sons Ltd, contractors, Gretna Munitions Factory track laying contract, Dumfries, by /1921. Locomotive may well have been used on the Branston Factory construction.

(1) to Trent Concrete Co Ltd, Colwick, Nottinghamshire.

CYCLOPS ENGINEERING CO LTD
VICTORIA STREET WORKS, Burton on Trent. SH18
(Company registered 5/1919)

A firm of engineers manufacturing tanks, drums, barrels, etc.. Sidings led into one of the firms' works from the Midland/LMS Horninglow branch. The company purchased the Baguley's works in 1932 and this is said to have used the Kay locomotive listed here. By the end of World War 2, part of this works had passed to the Electric Furnace Co, and the remainder to SBK Joinery.

Gauge: 4ft 8½in

	4wPM	Kay	c1930	New	s/s.

DARLASTON COAL & IRON CO LTD
ESSINGTON WOOD COLLIERY, Essington. SF1, SF11
Darlaston Steel & Iron Co Ltd until 8/1877
Samuel Mills until 10/1864

Colliery leased by Samuel Mills from the Vernon family who had worked the mines at an earlier date. Mr Mills, at the time was working the Darlaston Green Furnaces and other collieries in the Darlaston district. A railway was built to link the mines with a basin beside the Wyrley and Essington canal. It ran parallel to a disused canal branch made to serve the Essington mines and also replaced the earlier tramroad built to Newtown for the Vernon family (which see - section 7). A locomotive is said to have been used on this line in preference to rope haulage. In August 1875, another railway was constructed, this time to link up with the LNWR Walsall to Rugeley branch at what became known as Essington Wood Sidings.

Built by contractor Henry Lovatt, this line was worked by LNWR engines as far as Bursnip Lane Level Crossing, where a connection was made with the existing railway. The Darlaston Coal & Iron Co Ltd used its own locomotives to work the traffic from Essington Wood Colliery (SJ 965033) to the exchange sidings. The company also had a commitment to maintain part of the LNWR branch. Locomotives were based at a small shed near Bursnip Lane.

During 1883 plans were made to alter the surface workings at the mine. Automatic haulage arrangements were installed to replace the original system of delivery by locomotive allowing for the disposal of at least one of their locomotives. Additional seams of coal were proved in 1891 and a new company called the Holly Bank Colliery Co Ltd was created to take charge of all the plant and railways worked by the Darlaston Coal and Iron Co Ltd. For later history see under Hilton Main and Holly Bank Colliery Co Ltd.

Gauge: unknown
One locomotive, details unknown.

Gauge: 4ft 8½in

0-6-0ST	OC	FW	247	1874	New	(1)
0-6-0ST	OC	FW	286	1875	New	(2)

(1) to Nixons Navigation Co Ltd, Nixons Navigation Colliery, Merthyr Vale, Glamorgan, by /1903.
(2) to East Bristol Collieries Ltd, Kingswood Colliery, Bristol, Glos, by 6/1886. Possibly per J.Wilkes, Pelsall Foundry.

ESSINGTON FARM COLLIERY CO
ESSINGTON FARM COLLIERIES.   SF2

The Essington Farm Colliery Co first leased land for the colliery from the Vernon family 1/12/1873 and laid two short tramroads to the Wyrley and Essington Canal. These ran from two pits which the company sank on the property. In June 1893 the mineral property formerly worked by the Hilton Colliery Co (which see) was leased and this mine was reopened. About the same time other property was leased and a new mine called Springhill was sunk. The Springhill plant (SJ987039) was erected on a strip of land sandwiched between the LNWR Walsall to Rugeley line and the Wyrley and Essington Canal. It was served both by a standard gauge siding laid beside the screens and a canal basin built into the colliery site. Standard gauge sidings which connected with the LNWR Walsall to Rugeley line were completed in December 1892. A narrow gauge line was also laid in 1893 to connect Hilton (SJ981031) and the Springhill collieries. One locomotive was employed to work pit tubs between Hilton (SJ981031) and the Springhill mine. Colliery closed and plant auctioned 18/5/1904.

Gauge: 1ft 8in

    MARGARET   0-4-0ST   OC   WB   1429   1893 New (1)

(1) to G.Louis Lavender Ltd, Atherton Limeworks, Rushall, /1904.

THE FAIR OAK COLLIERY CO LTD
FAIR OAK COLLIERY, Near Rugeley.   SA4

Sinking operations at this pit were commenced on 1/1/1872 with the cutting of the first sod by Major Arden. Sinking proceeded slowly and problems were encountered during the operations, especially with water. Standard gauge sidings connecting with the LNWR Cannock Mineral line were completed 9/9/1872; a railway was laid to serve the No.1 plant (SK019163), but this pit never raised any coal.

By 1875 the shaft was 271 yds deep and headways were being made, yet all efforts proved unsuccessful and a second mine was started by 1878. Coal was eventually found, but proved to be workable for only a few years. A narrow gauge rope worked tramroad c500yds long was laid from the No.2 plant (SK016157) to No.1 to convey the pit tubs to a wharf beside the standard gauge siding. Both rope haulage and horses were employed underground. During 1879 a legal dispute developed concerning ownership of FW 382, which Fair Oak had acquired on 1st August 1878 under a hire purchase agreement with the Yorkshire Wagon Company. Following bankruptcy of Fox Walker in December 1878, Yorkshire Wagon sued Fair Oak for the monies outstanding on the purchase, and the matter was eventually resolved. During February 1884 attempts were made to prove deeper coal measures, but nothing was found. All the workforce was laid off from 9/7/1884 and the horses brought out of the mine. With the seams worked out, the company went into liquidation and this was announced during November 1884. Mr R.Mackay was appointed liquidator. The plant was disposed of by auction 24/25 and 26 November. One locomotive was mentioned in the sale and 65 wagons were bought by Cannock & Rugeley Colliery Co Ltd.

Gauge: 4ft 8½in

0-6-0ST   OC   FW   382   1878   New   (1)

(1) to Lidgett Colliery Co Ltd, Barnsley, Yorkshire (WR), c12/1884.

GIBBS & CANNING LTD
GLASCOTE TILERIES, Glascote, (Warwickshire until /1974)   SK2
Gibbs and Canning until 8/1878

These works (SK230032), manufacturing bricks and fireclay products, were established in 1847 by John Gibbs and Charles Canning. They were built close to the Glascote Colliery and served by a siding which connected with the Glascote and Amington Colliery Railway. Gibbs and Canning used the Glascote Colliery Railway to work its traffic to the Midland Railway exchange sidings at Kettlebrook. Part of their traffic was also conveyed along the Amington Colliery line by the Glascote Colliery Co (which see) to the LNWR. Gibbs and Canning Ltd was in liquidation 7/1881 and was suceeded by another firm of same name. Locomotives were employed on the line from the Tileries to Kettlebrook by 1881. The standard gauge system at the works was abandoned 6/1951 following the closure of the Amington Colliery railway and most of the track was removed by 7/1952.
The narrow gauge line was in existence by 1901, when it ran north and then east from the works to a clay mine (SK233033). It was later diverted to a quarry (SK 233036). Disused by 1954

Gauge: 4ft 8½in

GLASCOTE	0-4-0ST	OC	HE	282	1882	New	(1)
GLASCOTE No.1	0-4-0ST	OC	HE	467	1888	New	Scr /1917.
GLASCOTE No.2							
(GLASCOTE No.1)	0-4-0ST	OC	HE	22	1867	(a)	s/s
GLASCOTE	0-4-0ST	OC	YE	1027	1910	(b)	(2)
	0-4-0DM		KS	4428	1929	(c)	(3)
GLASCOTE No.2	0-4-0ST	OC	KS	4226	1930	(d)	(4)
	0-4-0ST	OC	P	832	1900	(e)	(5)

(a)   ex G.Wythes, contractor, BILLY, originally used on the East and West India Dock construction in London.
(b)   ex Thos.W.Ward Ltd, Sheffield, /1915.
(c)   ex KS, 12/1929.
(d)   ex KS, 5/1930, as new, though actually built in 1923.
(e)   ex Huntley & Palmers Ltd, Reading, Berks, via Geo Cohen Sons & Co Ltd, Kingsbury, Warwicks, /1949.

(1)   to T.A.Walker, Manchester Ship Canal construction, GREENHEYS.
(2)   to Newton Chambers & Co Ltd, Thorncliffe, Yorks (WR), 3/1952.
(3)   to KS 5/1930, then to HE, /1930. To Air Ministry, Cranwell, Lincs, 2/1933.
(4)   to HE for repair, 3/1950 but scrapped at HE, 4/1950.
(5)   to New Cransley Iron & Steel Co Ltd, Kettering, Northants, 9/1952.

Gauge: 2ft 6in

	4wWE	BE	16329	c1917	(a)	s/s c/1960.
	4wWE	BE	16330	c1917	(a)	s/s c/1960.

(a)   probably built for MoM use.

GLASCOTE COLLIERY CO LTD
GLASCOTE COLLIERY, Tamworth, (Warwickshire until /1974)   **SK4**
AMINGTON COLLIERY, Tamworth, (Warwickshire until /1974).   **SK3**
Glascote Colliery Co until 6/1911

Glascote Colliery was sunk about 1840 by Gibbs & Co and this mine was served by a tramroad built in two parts. The first was a level horse worked section which ran from the Birmingham and Derby Junction/Midland Railway near Kettlebrook as far as a canal basin on the Coventry Canal where it made a connection with the bottom of the second section, an incline up to Glascote Colliery worked by a stationary steam engine. Later this line was worked by Gibbs and Canning (which see).

In 1858 William Charles Firmstone and Henry Onions Firmstone leased the tramway and mines at Glascote for 40 years. The Firmstone family, who operated mines and furnaces in the Black Country (see West Midlands Handbook), worked these mines under the title Glascote Colliery Company.
Amington Colliery, also worked by the Firmstone family, under the title of the Amington Colliery Co, was sunk in 1862 and linked to a wharf beside the LNWR Trent Valley line at Alvecote known as Amington sidings. Later a line was laid to link both the Amington and

Glascote collieries passing by and providing a connection with the Terra Cotta works of Messrs Gibbs & Canning (which see).

Private owner wagons were delivered to this firm by 1865 and it is also likely that a locomotive was in use by this date, though its identity remains unknown. The locomotive shed was built at Amington Colliery (SK241036). Amington Colliery passed to NCB 1/1/1947.

Gauge: 4ft 8½in

1		0-4-0ST	OC	BE	317	1909	Scr by
	Reb	4wVBT	VCG	S	6661	1926 (a)	/1939
AMINGTON No.2		0-6-0ST	OC	HL	3642	1925 New	
	Reb			WB		1934	(1)
AMINGTON No.3		0-6-0ST	OC	WB	2508	1934 New	(1)

(a) probably acquired second hand.

(1) to NCB WM Div Area 4, 01/01/1947.

GREAT WYRLEY COLLIERY CO LTD.
GREAT WYRLEY COLLIERY, Cheslyn Hay.   SD1
Bernard Gilpin until 10/7/1875

Several mines were worked in the Great Wyrley area by the Gilpin family and these were linked to the Churchbridge edge tool manufactory by a tramway built in 1817. The Wyrley mines were operated with great success by several members of the Gilpin family even prior to Bernard Gilpin (see Non-locomotive lines section 7). With the development of the Cannock Chase Coalfield a deeper seam was believed to exist below the Wyrley measures and a limited company was promoted to prove and exploit these. This firm purchased the Churchbridge colliery, plant and tramways belonging to Bernard Gilpin. It also purchased the assets of the Great Wyrley Colliery Co Ltd, registered 13/06/1871 which mined an adjacent property of 60 acres but had neither rail nor canal connections. The new company, registered 10th July 1875, took the name of the Great Wyrley Colliery Co Ltd. and rapidly set about sinking a new mine shaft near to the LNWR Cannock - Walsall branch. The new sinking became known as the No.3 Pit (SJ986068). Coal continued to be mined at the No.1, 4 and 5 Pits of the former owners. The brickworks at No.2 Pit supplied bricks for the new sinking. The existence of a Cannock Chase seam was proved in 1877 and the making of permanent shafts commenced. Standard gauge sidings, completed by the LNWR on 8/3/1880, were also built to link the mine with the LNWR. The bulk of the company's traffic was sent by rail and the Great Wyrley Co Ltd set about buying a fleet of rail wagons through a hire purchase agreement.

By 1880 headings were being driven underground and the mine was finished in 1881. Both rope haulage and horses were employed in the workings underground. A sizeable portion of the coal continued to sent by canal or to the Churchbridge Ironworks, which was now in separate hands. Bernard Gilpin's tramways to the Churchbridge Ironworks and to the Wyrley Bank Canal remained intact. Two locomotives, purchased by Gilpin about 1875 to work the tramroad, had passed to the Great Wyrley Colliery Co Ltd by 1877. A least one appears to have been of DeWinton make and was the subject of a court action in 1880 when DeWinton attempted to sue Gilpin for non payment. In March 1880 the Great Wyrley Colliery Co Ltd applied to the Staffordshire and Worcester Canal Company to repair its Cheslyn Hay Tramroad (which see). The canal tramway from Walkmill Bridge appears to have been diverted about this time and ran directly to the Great Wyrley mine. There were now three narrow gauge tramways which converged at No.3 Plant. Further locomotives were purchased and they were employed to haul pit tubs to the canal wharves.

By 1915 another pit, worked by the Wyrley Cannock Colliery Co Ltd (which see -section 7) in earlier days, had been put back into production by the company. Located at the terminus of the Wyrley Bank Canal, it became known as the Nook or No.2 Plant. A tramroad was laid from its pithead to join the Gilpins arm tramroad.
The Great Wyrley Colliery Co closed Nook mine as uneconomic in 1924. In April 1925 the Great Wyrley Colliery also ceased working and the plant lay idle. During 1926 the property was acquired by the Nook & Wyrley Colliery Co, which see for later history.

Gauge: 2ft 0in

	4wTG					New	s/s.
	0-4-0T					New	s/s.
GORDON	0-4-0IST	OC	WB	1122	1888	New	s/s.
GREAT WYRLEY No.2	0-4-0ST	OC	MW	1246	1892	New	s/s.
COLONEL WILSON	0-4-0ST	OC	MW	1371	1897	New	
				Reb	1911		(1)

(1)   to Nook & Wyrley Colliery Co, with mine, 10/1926.

WILLIAM HARRISON LTD.
William Harrison until 10/1890.

BROWNHILLS COLLIERIES   SE17
CATHEDRAL PIT   SE12
GROVE COLLIERY   SE13
BROWNHILLS No.3 COLLIERY   SE15

Several coal mines were sunk on Brownhills Common by Harrison during the nineteenth century. The first pits bore the names Blue, Red and White, but later became known as Brownhills (or sometimes the Old Brownhills) Colliery. They were located near the Wyrley and Essington Canal at a place known as The Slough (SK 031051). Gradually a tramroad system developed to serve these pits.

By 1870 there were two main collieries in production; one was called Cathedral (SK027064 - worked by Harrison by 1854) and the other Wyrley Common (SK025059) (**SD18**). Both were linked to the LNWR Norton branch by a standard gauge mineral line. During the mid 1870's a new sinking was made at the Grove (SK018061) which stood beside the Cannock Extension canal. The private railway was extended across the canal to serve the mine.

At first the locomotives which worked these lines were stabled at a locomotive shed placed at the Cathedral Pit. In 1887 this colliery closed and the plant was sold. However locomotives continued to be based here until a new shed was erected at Grove Colliery. Wyrley Common Colliery closed about 1890. By 1897 another sinking was underway this time near the village of Great Wyrley. This mine was called Brownhills (later Wyrley) No.3 (SK002067). A 2ft 0in gauge main and tail rope tramroad ran from this pit to the Grove Colliery. Both Grove and Brownhills No.3 (by then known as Wyrley No.3) Collieries passed to the NCB WM Area 2, 1/1/1947.

Gauge: 4ft 8½in

1		0-4-0ST	OC	Lill		1864	New	s/s
	WARRIOR	0-6-0ST	IC	Lill		1867	New	Scr c/1934
	SUCCESS	0-6-0ST		JS		c/1869	New	Scr c/1913
	BLACK PRINCE	0-6-0ST	IC	RS	630	1848	(a)	(1)
	AGINCOURT	0-6-0ST	IC	RS	631	1848	(b)	Scr c/1906
	EMLYN	0-6-0ST	IC				(c)	Scr c/1920
No.3		0-6-0ST	IC	P	618	1895	New	(2)
	THE COLONEL	0-6-0ST	IC	HC	1073	1914	(d)	(2)

(a) ex LNWR, 1810, 10/1873. (orig LNWR 236, Reb from 0-6-0 IC, /1864).
(b) ex LNWR, 1811, 10/1873. (orig LNWR 241, Reb from 0-6-0 IC, /1864.
(c) ex C.D.Phillips, dealer, Newport, Mon, c/1905. Was said to be a double framed loco and possibly an ex-GWR 0-6-0ST.
(d) ordered by Houghton Main Colliery Co Ltd, Yorkshire on William Harrison's behalf. Probably delivered new to Brownhills.

(1) frame and other parts used as a winding engine at Rising Sun Colliery, Brownhills. The rest was scrapped on site, c/1909.
(2) to NCB, WM Division, Area 2, 1/1/1947 with colliery.

MID CANNOCK COLLIERY SD2

Colliery originally sunk by the Mid Cannock Colliery Co Ltd., at Rumer Hill on the site of older surface workings. The Mid Cannock Colliery Company commenced sinking without even issuing a propectus or canvassing for shareholders. The company was eventually registered in 1873 by which time sinking of the two shafts was well under way. By 1875 coal had been proved 136yds from the surface.
Standard gauge sidings were made into the colliery (SJ 987091) from the LNWR Cannock branch, but no record exists of any private locomotives being employed to shunt them. The company went into liquidation during 1882. Within the mine, horse-worked 1ft 9in gauge track was laid. The plant was auctioned between 31/7/1882 and 2/8/1882 and thereafter the mine lay idle.

William Harrison Ltd decided to reopen the pit and made a start in 1913. New screens were built and railway sidings were laid again to connect with the LNWR Cannock line in 1914. A narrow gauge tramroad was also made to connect the mine with the Cannock Extension canal. Again there is no record of any locomotives being used at the pit.
The colliery passed to NCB, WM Division, Area 2, 1/1/1947.

See also Non-locomotive lines section for details of limestone mines worked by Harrison.

HAUNCHWOOD-LEWIS BRICK & TILE LTD.
G.W.Lewis Tileries Ltd until 1969

ROSEMARY TILERIES, CHESLYN HAY WORKS   SD3

The tile works (SJ975075) at Cheslyn Hay were worked by Joseph Walker until 1897 when the lease was transferred to Lewis Tileries. These works were situated near to the Old Coppice colliery and by 1913 had a standard gauge siding which joined the colliery railway. Locomotives were used on the narrow gauge tramways in the quarry. These were laid to the base of a rope worked incline which hauled the wagons up to the crusher. Rail traffic ceased in 1974.

Gauge: 2ft 0in

	4wDM	RH	168833	1933	New	(1)
	4wDM	RH	175118	1935	New	Scr /1967
	4wDM	RH	187056	1937	(a)	(2)
	4wDM	RH	195868	1939	New	s/s.
	4wDM	RH	264242	1949	(b)	(2)
VEROLI	4wDM	RH	432664	1959	(c)	(3)
	4wDM	MR	8592	1940	(d)	(2)
	4wDM	MR	8681	1941	(e)	(4)
	4wDM	MR	21282	1959	(f)	(5)

(a) ex Essington Works, 1/1970.
(b) ex William Nock Ltd, Holly Lane Brickyard, Erdington, Birmingham.
(c) ex Taylor Woodrow contractors, Rheidol Hydro Electric Scheme, Merioneth.
(d) ex Essington Works, /1972.
(e) ex Essington Works, /1973.
(f) ex A.M.Keef, Bampton, Oxfordshire, hire, c7/1974.

(1) to Essington Works, /1970.
(2) to A.M.Keef, Bampton, Oxfordshire, 11/1973.
(3) to A.M.Keef, Bampton, Oxfordshire, 11/1973. To Narrow Gauge Centre of North Wales, Gwynedd, 26/7/1976.
(4) to A.M.Keef, Bampton, Oxfordshire, 6/1976.
(5) retn to A.M.Keef, Bampton, Oxfordshire, by 4/1975.

ROSEMARY TILERIES, ESSINGTON WORKS   SF3

These works (SJ968045) were originally operated by Davis and Co. In 1904 a standard gauge siding was made to connect with the LNWR Hollybank Colliery line and was worked by LNWR locomotives. Narrow gauge tramways were also laid in the quarries to the base of an incline where the clay was tipped from the trucks into a larger wagon which was hauled up a 4ft 0in gauge incline to the crusher.
Tramway closed 1/1975 and the works closed shortly afterwards.

Gauge: 2ft 0in

4wDM	RH	168833	1933	(a)	s/s c/1966
4wDM	RH	170197	1934	(b)	(1)
4wDM	RH	187056	1937	New	(2)

4wDM	RH	223737	1944	(c)	Scr c/1966
4wDM	MR	7170	1937	(d)	(3)
4wDM	MR	8681	1941	(e)	(4)
4wDM	MR	8592	1940	(f)	(5)
4wDM	MR	8882	1944	(g)	(3)

(a) ex Cheslyn Hay Works, /1970.
(b) ex Haunchwood Brick & Tile Co Ltd, Stockingford, Warwickshire.
(c) ex Railway Mine & Plantation Equipment Ltd, dealers, 4/1954.
(d) ex Haunchwood Brick & Tile Co Ltd, Stockingford, Warwickshire /1953.
(e) ex A.M.Keef, Bampton, Oxfordshire, /1972. Formerly J.& A. Jackson Ltd, Stockport, Cheshire.
(f) ex A.M.Keef, Bampton, Oxfordshire, c/1971. Formerly Flettons Ltd, Kings Dyke Brickworks, Cambridgeshire.
(g) ex A.M.Keef, Bampton, Oxfordshire, 11/1973.

(1) to Abelson. Later to Bovis,contractors.
(2) to Rosemary Tileries, Cheslyn Hay, 1/1970.
(3) to A.M.Keef, Bampton, Oxfordshire, 6/1976.
(4) to Rosemary Tileries, Cheslyn Hay, /1972.
(5) to Rosemary Tileries, Cheslyn Hay, /1973.

T.A.HAWKINS & SONS LTD
OLD COPPICE COLLIERY.   SD4
Joseph Hawkins & Sons until 7/1911

Mines known as Old Coppice Colliery (SJ972079). These were originally owned by Edward Sayers, but by 1869 Joseph Hawkins was proprietor. In 1856 he also commenced working the Lanehead Bridge Colliery near Walsall. The Lanehead Bridge mine was only worked for a short period and mining operations concentrated at Old Coppice. Deeper shafts were being sunk at Old Coppice in 1876 and by 1878 were complete. In May 1878 Joseph Hawkins applied to the Staffordshire and Worcestershire Canal Co to lay down a tramway over his land to the Canal. He also made an application in November 1878 to make a railway from his mine to the LNWR at Churchbridge. Both schemes were approved, but only the canal tramway was built. It ran to the existing basin at Walkmill Bridge served by the canal tramroad from Cheslyn Hay. A larger canal basin for Hawkins traffic was considered in December 1881. Construction went ahead and a new basin was completed for Joseph Hawkins own use in February 1883. A new shorter tramroad ran from pit shaft direct to the canal wharves. Joseph Hawkins continued to press for a standard gauge link to his mine. In 1900 a standard gauge mineral line, worked from the start by Hawkin's own locomotives, was finally built to connect the colliery with the LNWR Churchbridge mineral branch. The original connection to Churchbridge passed to the south of the Staffordshire & Worcestershire Canal's Farm Lodge Reservoir but was later rebuilt with a new line and more extensive sidings on land between this reservoir and the canal.

In addition to colliery traffic, Hawkins' locomotives also handled traffic for the adjacent Rosemary Tileries, which see. Hawkins also owned its own brickworks which lay to the west of the mine and was served by standard gauge sidings connected with the colliery railway.

At the colliery there was an extensive system of narrow gauge track (2700 yards) on the surface which connected the pitheads with a washery, three screens and a dirt mound. In the underground workings both endless rope and direct haulage were used extensively

The colliery and brickworks passed to NCB, WM Division, Area 2, 1/1/1947.

Gauge: 4ft 8½in

HAWKINS	0-6-0ST	OC	P		809	1900 New	(1)
SONS	0-6-0ST	IC	Lill			(a)	Scr c/1936
(formerley EMLYN)		Reb	C.D.Phillips				
TONY	0-6-0ST	OC	HL		3460	1921	(b) (1)

(a) ex C.D.Phillips, Dealer, Newport, Mon, /1914.
(b) to U.A.Ritson & Sons Ltd., Preston Colliery, North Shields, Northumberland, via Geo Cohen, Sons & Co Ltd, Newcastle upon Tyne, /1927.

(1) to NCB WM Division Area 2, 1/1/1947.

HILTON MAIN & HOLLY BANK COLLIERIES LTD
HILTON MAIN COLLIERY   SF7
HOLLY BANK COLLIERY   SF4
Holly Bank Coal Co Ltd until 24/6/1932
Holly Bank Colliery Co Ltd until 2/7/1910

The Holly Bank Colliery Co Ltd was formed 26/1/1891 to exploit the rich seams of coal discovered by the Darlaston Coal & Iron Co Ltd at Essington (which see). Utilising the old plant and railways of the Essington Wood Mine (**SF1**), the new plant was built adjacent to the old pit shafts and sidings laid to the screens then being built. A locomotive shed (SJ967032) was made beside the Holly Bank Mine replacing a smaller building near Bursnip Lane.

To exploit the recently discovered coal measures a mineral railway was projected to run from the colliery to the LNWR at Tipton through Coltham, Willenhall and Wednesbury, but it was never completed. Only a short section was laid from the mines to the Coltham canal basin at Short Heath (SJ973008) (**SF5**), coming into use c8/1900. The older canal tramroad to Essington Farm then ceased to be used. The wagons which ran on this section were specially adapted to hold pit tubs which were taken to the basin to be unloaded into the waiting boats.

With the sale of Essington Farm colliery in 1904, Holly Bank commenced to work the coal measures at Hilton Colliery formerly held under lease and worked by the Essington Farm company. Hilton Colliery itself, was reopened as Sneyd Colliery (SJ979028) (**SF6**) and standard gauge sidings laid to the screens from the Coltham canal line. This appears to have happened around 1906 and coincides with the acquisition of the locomotive NELLIE from G.R.Thomas.

Other railway schemes were proposed. In 11/1898 a group of Hollybank directors applied to the Light Railway Commissioners to build a standard gauge light railway, the Essington & Ashmore Light Railway, to connect Hollybank Colliery to the Midland Railway (Walsall to Wolverhampton line) at Bentley and the LNWR at Darlaston. There would have been a short branch to a new basin by the Wyrley & Essington Canal at Wednesfield. However the LNWR objected to the proposals and the line was not built.

During 1914, the Holly Bank concern sought to exploit fresh coal measures and searched on the west side of the South Staffordshire boundary fault. However due to the outbreak of the

First World War further explorations were curtailed and it was not until after the war that they recommenced. The company found coal and in December 1919 started sinking the Hilton Main colliery (SJ941043) (**SF7**), north of Wolverhampton. A new railway was then built by a contractor between 1922 and 1924 to link the mine with the existing system at Holly Bank. Hilton Main was completed in 1923 but not formally opened until September 1924. A portion of this line was on the route of the projected Wolverhampton & Cannock Chase Light Railway (see Section 7). A paddy train which took miners to and from work ran from Short Heath Coal Wharf for the full length of the line to Hilton Main. The train comprised passenger vehicles, which were simply coal wagons with a covered roof, and a brakevan at the rear. Holly Bank and Hilton were also connected through underground workings. Sneyd mine closed completely in 1920 and was dismantled, while Holly Bank ceased coal winding in 1927 after Hilton opened, but continued to be used for pumping operations. Locomotives continued to be based at Holly Bank shed, though an engine siding was provided at Hilton Main. By 1944 limited coal working had resumed at Holly Bank (No.3, 5, 7 and 15 Pits) but this is thought to have ceased again by Vesting Day.

During the early 1930's the firm fell into financial difficulties and by 1932 the company was in the hands of the liquidators. The concern survived through the efforts of Mr C. Nelson, chairman and managing director of the Hartley Main Collieries Ltd, Cramlington, Northumberland, who in 1935 formed a new company to work the mines. Underground haulage was basically by horses at Holly Bank and endless rope powered by electricity at Hilton Main.

Both Hilton Main and Hollybank passed to the NCB, WM Division, Area 2, 1/1/1947.

Gauge: 4ft 8½in

HOLLY BANK No.1	0-6-0ST	IC	HC	353	1893	New	Scr /1934.
HOLLY BANK No.2	0-6-0ST	IC	HC	568	1900	New	s/s.
NELLIE	0-6-0ST	IC	MW	563	1876	(a)	Scr.
HOLLY BANK No.3	0-6-0ST	IC	HE	1451	1924	New	(1)
HILTON MAIN No.10	0-6-0ST	IC	RWH	1665	1876		
	Reb		RS		1933	(b)	(2)
No.1	0-6-0T	IC	HC	352	1891	(c)	(1)
ROBERT NELSON No.4	0-6-0ST	IC	HE	1800	1936	New	(1)
CAROL ANN No.5	0-6-0ST	IC	HE	1821	1936	New	(1)

(a) ex G & R.Thomas Ltd, Hatherton Furnaces, Bloxwich, West Midlands, c/1906.
(b) ex Hartley Main Collieries Ltd, Northumberland (No.10), 10/1934. Originally NER 1358.
(c) ex Hartley Main Collieries Ltd, Northumberland (No.23), /1935. Originally Barry Railway 53, and later GWR 785.

(1) to NCB, WM Division, Area 2, 1/1/1947.
(2) to Moira Colliery Co Ltd, Leicestershire, 1/1943.

IND COOPE & CO LTD.
BURTON ON TRENT BREWERY. SH9
Ind Coope & Co (1912) Ltd until 1/1/1923
Ind Coope & Co Ltd until 21/10/1912
Ind Coope & Co until 11/1886

Messrs Ind Coope's brewery was established in 1856 and stood in Station Street (SK244233) with maltings in Mosley Street. A branch (the Mosley Street branch) was made from the Midland Railway south-west of Burton station to the Maltings and across Station

Street into the Brewery yard (opened 13/3/1865). The company's first locomotive is known to have commenced delivering traffic to the Midland Railway on 17/10/1867 and subsequently two locomotives were retained to handle this traffic. A locomotive shed was built close to the maltings (SK242230). The firm was amalgamated with the adjacent brewery of Samuel Allsopp & Sons Ltd in 6/1934 and the locomotive fleets were then merged.

Gauge: 4ft 8½in

		0-4-0WT	OC	TW?		1867 New?	s/s.
No.1		0-4-0ST	OC	RWH	2022	1885 New	Scr.
No.2		0-4-0ST	OC	HL	2295	1895 New	(1)
(No.3)	No.2	0-4-0ST	OC	HL	2345	1896 New	
		Reb	TW			1925	(2)
No.1		0-4-0ST	OC	HL	3539	1923 New	(3)
No.3		0-4-0ST	OC	HL	3632	1925 New	(3)

(1) to Synthetic Ammonia and Nitrates Ltd, Billingham, Co Durham, via H.Young, Darlington 9/1925.
(2) to Romford Brewery, Essex, /1925. Retn to Burton. To Ind Coope & Allsopp with works, 6/1934.
(3) to Ind Coope & Allsopp Ltd, with premises, 6/1934.

IND COOPE LTD.
BURTON ON TRENT BREWERIES

ALLSOPP'S OLD BREWERY SH24
NEW BREWERY SH1
IND COOPE BREWERY SH9

Ind Coope & Allsopp Ltd until 31/12/1958
(Member of the Allied Breweries Group)

In June 1934 the two brewery firms of Ind Coope Ltd and Samuel Allsopp & Sons Ltd merged. Locomotives working for the two breweries were numbered as one fleet, but the two locomotive sheds were retained at the separate breweries. Allsopps shed closed in 1967 while the former Ind Coope shed continued in use until 1970 when rail traffic ceased.

Gauge: 4ft 8½in

No.1	0-4-0ST	OC	HL	3539	1923	(a)	(1)
No.2	0-4-0ST	OC	HL	2345	1896	(a)	Scr 6/1949
No.3	0-4-0ST	OC	HL	3632	1925	(a)	(2)
No.4	0-4-0ST	OC	HL	3540	1923	(b)	(3)
	0-4-0ST	OC	HCR	178	1876		
	Reb	TW			1901	(c)	s/s c/1934
No.6	0-4-0ST	OC	AB	1858	1925	(c)	(4)
No.7	4wVBT	VCG	S	9376	1947	New	(5)
No.8	4wVBT	VCG	S	9384	1948	New	(6)
No.9	4wBE		EE	533	1922	(d)	s/s c/1968
No.1	0-4-0DM		Bg	3357	1952	New	(7)
No.2	0-4-0DM		Bg	3227	1951	(e)	(8)

(a) ex Ind Coope & Co Ltd, 6/1934.
(b) ex Ind Coope & Allsopp Ltd, Romford Brewery, Essex,/1937.

South Staffordshire Handbook. Page 68

(c) ex Samuel Allsopp & Sons Ltd, 6/1934.
(d) ex WD, Bramley, Hampshire, 11/1946.
(e) ex Romford Works, Essex, /1955. Incorporates parts of HL 3632.

(1) to Romford Works, Essex, c/1948, retn by 9/1951. To Thos. W. Ward Ltd, Templeborough Works, Sheffield, c/1952, then to NCB Avenue Carbonisation Plant, Wingerworth, Derbyshire, /1953.
(2) parts incorporated into Bg 3227 0-4-0DM and sent 'new' to Romford Works, Essex, /1951. The rest was scrapped.
(3) to Thos. W. Ward Ltd, Templeborough Works, Sheffield, /1953. To Low Moor Best Yorkshire Iron Co Ltd, Bradford, /1953.
(4) Scrapped /1954, after 2/1954
(5) to Thomas Hill (Rotherham) Ltd., Kilnhurst, Yorks (WR), /1960. To East Ardsley Colliery, Yorks (WR), c7/1960.
(6) scrapped on site 14/1/1971.
(7) to Derbyshire Stone Ltd, Wirksworth Quarries, Derbyshire, 3/1970.
(8) to Derbyshire Stone Ltd, Cawdor Quarries, Derbyshire, 3/1970.

KETTLEBROOK COLLIERY CO LTD
KETTLEBROOK COLLIERY, Wilnecote, (Warwickshire until /1974).   SK5
J and T Dumolo until/1881
Edward Best until c/1850

Edward Best is recorded in 1818 as the proprietor of the Park & Kettlebrook Collieries. Kettlebrook Colliery (SK227025) was linked through a tramroad (SK9) to a wharf and basin beside the Coventry Canal at Kettlebrook. This tramroad, originally horse worked, was shown on maps which date from 1834 but was probably built earlier. By 1854 John Dumolo had become proprietor of the Kettlebrook Colliery. At a later date the tramroad was relaid along a different course when it ran to a new basin beside the Coventry Canal (SK5) and also crossed the canal to reach a wharf next to a siding of the Midland Railway. One locomotive is known to have been used by Dumolo. It was advertised for sale by auction 23/1/1880 along with 28 coal wagons and pit ponies. The mine was subsequently worked by the Kettlebrook Colliery Co Ltd which replaced locomotive haulage with rope haulage on the canal tramroad. Colliery closed 1886. A second shaft known as Dumolo's Colliery was located SE of Kettlebrook at SK 229021 and may have been served by the tramroad.

Gauge: 2ft 7¾in

                    0-4-0T   OC   HH   c/1875 New? (1)

(1) For sale 23/1/1880; s/s

LITTLEWORTH EXTENSION RAILWAY   SC11

This short section of line was built between 1880 and 1881 for the LNWR by the contractor H.Lovatt. It linked the Norton branch with the Birmingham Canal Navigation Littleworth tramroad. Worked by Cannock Chase Colliery locomotives, it provided a shorter route to the LNWR than hitherto existed for coal drawn at Hednesford Colliery.

THE LITTLETON COLLIERIES LTD.

CONDUIT COLLIERY, Norton Canes.   SE7

Two mines (No.3 and No.4) were acquired in 1931 from the Conduit Colliery Co Ltd, (which see). No.4, also known as Norton Green, was subsequently closed (after 1945), while No.3 remained in use to be taken over by the NCB WM Div No.2 Area, 1/1/1947.

Gauge: 4ft 8½in

CONDUIT No.1	0-6-0ST	IC	MW	244	1867	(a)	(1)
CONDUIT No.3	0-6-0ST	IC	MW	1180	1890		
		Reb	MW		1920	(a)	(2)
CONDUIT No.4	0-6-0ST	IC	MW	1326	1896	(a)	(1)
AYNHO	0-6-0ST	IC	MW	1722	1908	(b)	(1)

(a)   ex Conduit Colliery Co Ltd with colliery, /1931.
(b)   ex Walter Scott and Middleton, contractors.

(1)   to NCB WM Division, Area 2, 1/1/1947 with the mine.
(2)   to Littleton Colliery, Huntington.

LITTLETON COLLIERY, Huntington.   SB2, SB4

The colliery (SJ972128) was commenced in 1872 by the Cannock & Huntington Colliery Co (see Non-locomotive lines section) but was never completed due to water flooding the shafts in July 1881. Work was recommenced in 1897 by Lord Hatherton and a new company, The Littleton Collieries Ltd, was registered 10/6/1899 to work the mine. The chairman was Sir G.H Holcroft of the Conduit Colliery Company. This new work proceeded rapidly and the sinking of the No.2 shaft was restarted on 6/10/1897. This sinking was completed to a depth of 1642 feet by February 1899. During July 1900 work then started on the No.3 shaft and this had been sunk to a depth of 1663 ft by 22/11/1902.

A standard gauge railway was made along the course of the original horse tramroad from the Cannock and Huntington mine to the canal at Otherton. This was on a straight course but the derelict tramroad had to be widened and the existing bridges rebuilt. A new section was then laid, which crossed the Staffordshire and Worcester canal and carried on to join the LNWR Wolverhampton to Stafford line at Penkridge. At Otherton several sidings were laid, two finishing beside the quayside while a third terminated on a high level gantry built in the centre of a large canal basin. Construction of the railway took about two years and the traffic for the colliery was initially handled by the contractors, Messrs A. Braithwaite. The Littleton Colliery's own locomotive commenced working the railway from 1901.

A large proportion of the traffic was taken along the full length of the line to the Penkridge exchange sidings but some was sent away by canal. Cobbles and large coal destined for the canal were screened at the pits and loaded into 2-ton capacity boxes which were conveyed to the canal basin on flat railway truck frames (three boxes per flat). At the basin the boxes were loaded into boats by a rail steam crane. Small nuts or coal slack for transhipment by canal were simply tipped down shoots from the high level gantry at the basin into waiting boats.

Surface narrow gauge lines were restricted to the stockyard and the pithead area. Underground lines were narrow gauge and haulage was usually performed through endless ropes driven by electric motors though some horses were also used in the workings.

Colliery passed to NCB, WM Division, Area 2, 1/1/1947.
Ref: Sinking and Equipment of the Littleton Collieries- T.H.Bailey.

Gauge: 4ft 8½in

	LITTLETON No.1	0-6-0ST	IC	MW	1515	1901	New	(1)
	LITTLETON No.2	0-6-0ST	IC	MW	1596	1903	New	(1)
	LITTLETON No.4	0-6-0ST	IC	MW	1759	1910	New	(1)
	LITTLETON No.5	0-6-0ST	IC	MW	2018	1922	New	(1)
	CONDUIT No.3	0-6-0ST	IC	MW	1180	1890		
			Reb	MW		1920	(a)	(1)
6	ADJUTANT	0-6-0ST	IC	MW	1913	1917	(b)	(1)

(a) ex Conduit Colliery, Norton Canes.
(b) ex Cannock & Rugeley Colliery Co Ltd, loan 9/1945.
(1) to NCB, WM Division, Area 2, 1/1/1947, with mine.

MW 1515 ran for a period in 1946 as 'Littleton No.2'.

LLOYDS (BURTON) LTD
WELLINGTON WORKS, Burton on Trent. SH10
A subsidiary, registered 10/1947, of F.H.Lloyd & Co Ltd.

Iron and steel foundry (SK236226) established by the firm of F.H.Lloyd. They were operated as MoS Burton during World War II and the railways sidings appear to date from this time. Rail traffic ceased c/1970.

Gauge: 4ft 8½in

		4wDM	RH	218045	1942	(a)	(1)
		4wDM	RH	386873	1955	New	s/s c/1970

(a) New to MoS Burton, 12/1942 to the order of H.A.Brassert & Co Ltd, Consulting Engineers.

(1) to C J Driver Ltd, Queslett Rd, Great Barr, Birmingham, by 21/5/1957; later to Abelsons & Co (Engineers) Ltd, Sheldon, Birmingham, 10/1957. Resold to John Bagnall & Sons Ltd, Leabrook Ironworks, Wednesbury, (see West Midlands Handbook), 5/1965.

WILLIAM LOUNDS, COLLIERY CONTRACTOR
CANNOCK LODGE COLLIERY, Bloxwich. SF8
Sarah Lounds until/1895

The Cannock Lodge Colliery Co Ltd leased 415 acres of mines at Bloxwich and its sinking reached coal in March 1877. A railway, completed 3/1880, connected the Cannock Lodge No.1 Colliery (SJ988045) with the LNWR at Norton Cannock Colliery sidings, completed 8/1875. A short branch of the Wyrley Bank canal was also built towards the mine. Traffic on the railway was worked by Sarah Lounds. In 1895, upon Sarah Lounds death, the

South Staffordshire Handbook. Page 71

locomotives were put up for sale. Later Peckett of Bristol delivered a new locomotive to W. Lounds at Cannock Lodge, but it probably saw little use here. Cannock Lodge Colliery Co ceased coal winding in 1895 and went into liquidation in 1896. The plant was advertised for sale 7/1896, while the mine was purchased by the Norton Cannock Coal Co Ltd.

Gauge: 4ft 8½in

0-4-0ST	OC	HCR			(a)	s/s c/1895
0-4-0ST	OC	P	440	1885	(b)	(1)
0-4-0ST	OC	P	597	1895	New	(2)

(a) origin unknown.
(b) ex Cadbury Brothers Ltd, Bournville, Worcestershire, /1894.

(1) to Buggins & Co Ltd, dealers, Birmingham. Resold about 1896 to Price and Wills, contractors, TEVERSAL.
(2) to Walsall Corporation, Pleck Gasworks, 12/1896.

SIR ALFRED Mc ALPINE
MILFORD GRAVEL PITS, Brocton SA5

Tramway laid in quarry.

Gauge: 2ft 0in

4wPM	MR	7939	1939	New	(a)

(1) to Tarmac, Blue Rock Quarries, Oldbury by 6/1944.

MARSTON, THOMPSON & EVERSHED LTD
BURTON ON TRENT BREWERY. SH11
J Marston, Thompson and Son Ltd, registered /1898, until/1905
J Marston & Son Ltd, registered 6/1896

John Marston was an established brewer at Horninglow, Burton before 1750. In 1898, the firm of John Marston & Son Ltd merged with that of J & F Thompson, whose brewery stood in Horninglow Street, to create the firm of J Marston, Thompson and Son Ltd. In 1898 this company acquired the Albion Brewery, Shobnall, of Mann, Crossman and Paulin (see Non-locomotive lines section), a company which had moved to London. Mann's Brewery (SK231203) was already linked to the Midland Railway Shobnall branch and from 1901 the company's loco hauled traffic to the exchange sidings. After taking over the plant Marston and Thompson made various improvements and enlargements including the laying of extra rail sidings and commenced to use its own locomotive for internal haulage. The brewing business of Sydney Evershed Ltd, founded in 1854 and registered in 3/1889, (with own sidings into the Midland Railway's Bond End branch) was acquired in 1905 and thereafter the company was known as Marston, Thompson & Evershed. Brewing was concentrated at the Shobnall Brewery, but parts of the older constituent breweries were retained as well as their sidings with the Midland Railway:

Cold Storage, Horninglow Street- Linked with Midland Railway Saunders branch.
Horninglow Street Maltings- Linked with Midland Railway Guild Street Branch.
Park Street Maltings- Linked with Midland Railway Bond End branch.

Rail traffic ceased 6/1964 and connection with BR removed. Remaining locomotives, until disposal, were stored on the premises.
Bg 3410 returned to Brewery c/1990 for preservation and presently stands in the Bottle Yard.

Gauge: 4ft 8½in

No.1	0-4-0ST	OC	HL	2502	1901	New	(1)
No.2	0-4-0ST	OC	HL	2837	1910	(a)	(2)
No.3	0-4-0ST	OC	HL	3581	1924	New	(3)
No.4	0-4-0ST	OC	HL	3774	1931	New	Scr 2/1955
No.4	0-4-0DM		Bg	3410	1955	(b)	

(a) ex Herbertsons Ltd, Chollerford, Northumberland, c4/1919.
(b) New ex Baguley, but incorporates the frame of HL 3774: to Bristol Mechanised Coal Co Ltd, Filton, Gloucestershire c12/1966: retn by 3/1990 for preservation.

(1) to Thos.W.Ward Ltd, Sheffield; resold to Ford Motor Co Ltd., Dagenham, Essex, /1931.
(2) to HL, then resold to Priestman Collieries Ltd, Norwood Coke Ovens, Co Durham, 25/10/1932.
(3) to Foxfield Light Railway Society, Dilhorne, Staffs, 12/04/1967.

MINISTRY OF DEFENCE
BRANSTON CENTRAL ORDNANCE DEPOT, Burton on Trent.   SH4
War Department until 1/4/1964

The buildings (SK233215) and sidings previously used by Branston Artificial Silk Co Ltd (which see) were taken over by the WD in 1937 and extended. Rail traffic ceased c/1976.

Gauge: 4ft 8½in

		0-4-0PM	BgE	1654	1928	(a)	(1)
BRANSTON No.2	71681						
	WD 847	0-4-0DM	HE	2065	1940	New	(2)
BRANSTON No.3	71682						
	WD 848	0-4-0DM	HE	2066	1940	New	(3)
202	WD 823	0-4-0DM	AB	357	1941	(b)	(4)
	WD 850	0-4-0DM	HE	2068	1940	(c)	(5)
212	WD 863	0-4-0DM	DC	2169	1942		
			VF	4861	1942	(d)	(6)
247	WD 867	0-4-0DM	AB	342	1940	(e)	(7)
248		0-4-0DM	AB	344	1941	(f)	(8)

(a) ex Branston Artificial Silk Co Ltd, with premises, /1937.
(b) ex Bicester Depot (after 3/1955).
(c) ex Ruddington Depot, Ruddington, Notts, 2/12/1964.
(d) ex Fauld Depot, Staffordshire, 3/9/1968.
(e) ex Sudbury Depot, Staffordshire, 3/2/1969.
(f) ex REME Supply depot, Aldershot, 4/1/1973.

(1) to ROF Rotherwas, Herefordshire, c/1944.
(2) to Bicester Base Ordnance Depot, 15/2/1957, retn 20/9/1957. To Bicester, 28/1/1966. Returned from Sinfin Depot, Derbyshire, 3/8/1967. Sold to Stewart (Metals) Ltd, Stretford, Manchester for scrap 14/12/1973; engine removed and remainder scrapped on site /1974.
(3) to Bicester 13/8/1955; retn 15/2/1957; to Bicester 3/2/1965; retd 28/1/1966; To Bicester, 29/2/1972.
(4) to Cairnryan Military Railway, near Stranrear, Wigtownshire, 16/9/1957. Retn from Bicester, 25/2/1972. To 322 Engineer Park, Hessay, North Yorkshire, 12/3/1975.
(5) to Bicester, 18/9/1967.
(6) to Bicester, 26/5/1972.
(7) to Bicester, 4/1/1973.
(8) to Chilwell Depot, Notts, 10/3/1975

MINISTRY OF MUNITIONS
NATIONAL MACHINE GUN FACTORY, Branston.　　　　　　　　　　　　　　SH4

Work on this factory (SK233215) was commenced in 1918 by T.Lowe, contractor, and completed in 1919. Railway sidings were laid to connect with Midland Railway at Branston. The factory did not come into full production due to the ending of the War. It was sold to Crosse & Blackwell c5/1920 (which see). Locomotives used on its construction are believed to include:

Gauge: 4ft 8½in

　　　　　　　　　　0-4-0WT   OC   H(L)　　366   1866   (a)   s/s

(a)   ex another MoM contract.

Another locomotive may well have been HCR 139 which worked for Crosse & Blackwell after it purchased the works.

NEW CANNOCK & WIMBLEBURY COLLIERY CO LTD.
WIMBLEBURY COLLIERY　　　　　　　　　　　　　　　　　　　　　　　　SC5
Registered 9/5/1881.

Company commenced working the Wimblebury Colliery (SK014117) after it had been given up by the Cannock & Wimblebury Colliery Co Ltd, (which see). It worked the mine for only a few years, going into liquidation in 1887. Plant and locomotive were advertised for sale 11-13th July 1887. About 1889 the mine was reopened by the Cannock & Rugeley Colliery Co Ltd which then provided its own locomotives to handle the traffic.

Gauge: 4ft 8½in

　　　　　　　　　　0-6-0ST OC　　FW　　370   1878   (a)   (1)

(a)   believed delivered new to mine, c/1881.

(1)   s/s, or possibly to CRC with mine and disposed of soon after.

South Staffordshire Handbook. Page 74

NOOK & WYRLEY COLLIERY CO LTD
GREAT WYRLEY COLLIERY   SD1

This company was formed in October 1926 to take over the collieries formerly worked by the Great Wyrley Colliery Co Ltd, which see. Plant included the No.2 or Nook Mine and the No.3 or Great Wyrley Colliery (SJ 986068) but only the Great Wyrley Colliery was worked. The firm acquired the 2ft gauge line to the Wyrley Bank Canal but may not have used it. It did however continue to use the standard gauge siding. It had no standard gauge locomotive but a fixed steam engine provided power as necessary. Underground haulage was by endless rope and horses were also used extensively within the workings. The No.3 colliery passed to the NCB, WM Division, Area 2, 1/1/1947.

Gauge: 2ft 0in

COLONEL WILSON	0-4-0ST	OC	MW	1371	1897		
			Reb		1911	(a)	Scr /1944

NORTON CANNOCK COAL CO LTD.
NORTON CANNOCK COLLIERY, Bloxwich   SF9
Norton Cannock Colliery Co Ltd until 30/7/1877

The Norton Cannock Colliery Co Ltd was formed in 1874 and mining operations were underway by June 1875. There were two pits, a No.1 trial shaft and the No 2 pit. The No.2 colliery (SJ987045) was linked to LNWR Cannock Branch by a short mineral railway worked by Norton Cannock's own locomotives. During 1876, the Norton Cannock Colliery Co Ltd went into liquidation and the sidings closed. The mine and assets were taken over by the Norton Cannock Coal Co Ltd (registered 30/7/1877) and the mines restarted. From 1896 traffic for and from the Cannock Lodge Colliery (see W.Lounds) was also handled. Mines closed in 1910 and locomotives were offered for sale 7/10/1910.

Gauge: 4ft 8½in

CANNOCK No.1	0-6-0ST	IC	MW	556	1875	New	(1)
	0-6-0T	IC	MW	167	1865	(a)	(2)

(a)   ex ? ; originally Potteries, Shrewsbury and North Wales Railway, until c/1873

(1)   to Holwell Iron Co Ltd, Melton Mowbray, Leicestershire, c/1910.
(2)   to H.Speakman, Bedford Colliery, Leigh, Lancashire, c/1910.

PERRENS & HARRISON
WILNECOTE COLLIERY, Wilnecote, (Warwickshire until /1974).   SK6

This mine (SK225018) was established by 1855. A branch, in place by 1858, from Midland Railway Birmingham-Tamworth line (where a signal box "Perrin & Harrison's Sidings" remained until 1969) served Wilnecote Colliery and Brickworks (SK224021). In 6/1858 the Midland Railway allowed Perrens & Harrison to lay a tramway through one of the arches of the railway bridge which spanned Kettlebrook Lane. Dumolo's Tramway (which see) appears also to have passed under this bridge and therefore Perrens & Harrison may have laid a tramway to the Coventry Canal. Late in 1865 the Midland Railway relaid part of the sidings with heavier rail to cope with increased traffic and subsequently maintained them under an agreement of 1868. Colliery closed 1879 and line later removed. Although locomotives are presumed to have worked the branch this is not confirmed. Neither is it certain that the whole line was standard gauge.

RIVER TRENT CATCHMENT BOARD
ELFORD WORKSHOPS, Elford, near Lichfield SG2

Essentially the Board's work of maintaining river banks falls within Derbyshire, however some of this work was carried on around Staffordshire at Hopwas Bridge and the locomotives involved on this work are listed below. For fuller details see Derbyshire handbook.

Gauge: 2ft 0in

	4wDM	MR	5819	1935	New	s/s
	4wDM	MR	5825	1936	New	s/s
	4wDM	MR	5828	1936	New	s/s
	4wDM	MR	5829	1936	New	s/s

ROM LTD
TRENT VALLEY WORKS, Lichfield. SG3
Rom River Plasclip Ltd until 2/1985
Rom River Reinforcement Co Ltd, until 21/11/1983
Subsidiary of the Rugby Portland Cement Co Ltd.

Sidings built in 1963, linking with BR Rugby-Stafford line north of Lichfield Station (SK133102), to serve a new steel reinforcement stockholding factory. Rail traffic ceased by 1/3/1988.

Gauge: 4ft 8½in

	4wDM	FH	3809	1963	New	(1)
	6wDM	KS	4421	1929	(a)	(2)
No.12	0-4-0DH	NBQ	27940	1959	(b)	(3)
B308	0-4-0DH	NBQ	27814	1958	(c)	(3)

(a) ex NCB South Durham Area, Deaf Hill Colliery, 6/2/1968 (orig Ravenglass & Eskdale Rly, Cumberland). To Witham Works, Essex, 6/6/1979. Retn by 3/1982.
(b) ex Cadbury Schweppes Ltd, Bournville, West Midlands, 27/10/1976.
(c) ex Stewarts Road Works, Battersea, London, 11/1986.

(1) to Mr Taroni and presented to Lichfield Council for display in Beacon Park 10/1973.
(2) to Foxfield Light Railway Society, Dilhorne, N. Staffordshire, 2/11/1985
(3) to White Wagtail Ltd, Gun Range Farm Scrapyard, Shilton, nr Coventry, 10/8/1988

THOMAS SALT & CO LTD.
BURTON ON TRENT BREWERY. SH12
COOPERAGE SH26
Thomas Salt and Co until 11/1893

The brewery (SK233232), established c/1807-1812, was located in High Street, Burton. The cooperage, maltings and stores stood in Anderstaff Lane. The company started to use its own locomotives 25/11/1869 and from 1/1/1870 worked its traffic over the Hay Branch to Horninglow. An engine shed (SK254236) was built opposite the cooperage in Jan 1871 and was connected with the Midland Railway Hay Branch. The company went into voluntary liquidation 10/5/1928 and the business was acquired by Bass, Ratcliff & Gretton Ltd in 1927; brewery subsequently closed. Salt's loco shed was in occassional use by Worthington locos for many years afterwards.

Gauge: 4ft 8½in

		0-4-0WT	OC	TW		1869	New	(1)
		0-4-0ST	OC	JF	1573	1873	New	s/s
1		0-4-0ST	OC	HC	272	1884	New	
			Reb	HC		1923		(2)
2		0-4-0ST	OC	HC	576	1900	New	(3)
3		0-4-0ST	OC	BgC	2001	1920	(a)	(4)
	SWANSEA	0-4-0ST	OC	HC	724	1905		
			Reb	HC		1919	(b)	(5)

(a) built on chassis of 0-4-0PH BgC 614/1921, which had never been completed. Carries worksplate BgC 621/1919.
(b) ex HC, hire ; (orig. Swansea Harbour Trust).

(1) advertised for sale in the Colliery Guardian 24/2/1888, s/s.
(2) to Worthington & Co Ltd, /1928.
(3) to Wellingborough Iron Co Ltd, Northants, via Geo Cohen Sons & Co Ltd, /1928.
(4) to Greaves Bull & Lakin Ltd, Harbury, Warwickshire, via Geo Cohen Sons & Co Ltd, /1928.
(5) retd to HC after hire, 3/1920; later Semet-Solvay & Piette Coke Oven Co Ltd, Parkgate, Yorks (WR), c/1928.

GEORGE SKEY & CO LTD.
George Skey until 10/1863

TAME VALLEY COLLIERY, BRICK & TILE WORKS, Near Wilnecote, SK7
(Warwickshire until /1974).

George Skey purchased these works c/1869, prior to which they had been worked by the Tame Valley Company (which see in Section 7). Tame Valley Colliery (SK223004) and Works were connected by a short branch railway laid from the Midland Railway Birmingham to Derby railway. It is believed that locomotives (details unknown) were used at the Colliery prior to those listed below. Colliery closed c/1931 and locomotive usage ceased but rail traffic remained to the Brick & Tile Works, which was taken over c/1938 by Doulton & Co Ltd who used rail traffic until the late 1950's.

Gauge: 4ft 8½in

NELLIE	0-4-0ST	OC	P	937	1902 New	(1)
NELLIE	0-4-0ST	OC	P	1666	1924 New	(2)

(1) to P, /1925, then to Blackpool Corporation Gas Department, Lancs, /1925.
(2) to Hawfield Brick & Pipe Works, Newhall, Swadlincote, Derbyshire, /1931.

WILNECOTE PIPEWORKS and PEEL COLLIERY   SK8 , SK11

Sidings were built by the Midland Railway in 1860 which ran from the pipeworks (SK219016) to the Birmingham - Derby line north of Wilnecote Station. Narrow gauge tramroads ran east from the Wilnecote Pipeworks to Peel Colliery (SK223017). This mine had been sunk by 1859 and for a time was operated under the title of the Peel Colliery Company, controlled by George Skey. Another tramroad ran south to the Watling Street at SK223014. Peel Colliery was closed after 1904. Beauchamp and Wilnecote (new) Collieries were also served. The works were taken over by Doulton & Co Ltd c1938 and rail traffic ceased in the 1950's

TRUMAN, HANBURY, BUXTON & CO LTD.
BLACK EAGLE BREWERY, BURTON ON TRENT.   SH13
Registered 1/1889

In 1865 the Phillips Brothers established their brewery (SK244236) fronting Derby Street, with sidings authorised by the Midland Railway from 2/1865. In 1873 the London firm of Truman, Hanbury and Buxton took over the brewery extending the premises and laying down rail sidings throughout the site. These sidings connected with the Midland Railway Burton to Derby line which ran beside the brewery. The first locomotive (details unknown) started to work here 1/7/1880, and traffic continued to be worked by firm's own locomotives until 1964. A road tractor was used to work the remaining traffic with one locomotive remaining in reserve until 1965. Rail traffic ceased. Business acquired by Grand Metropolitan Hotels in August 1971. Brewery closed 1971 and demolished.

Gauge: 4ft 8½in

	0-4-0ST				(a)	s/s
	0-4-0ST	OC	YE	406	1886 New	s/s
NEWCASTLE	0-4-0ST	OC	HL	2507	1901 New	Scr 4/1954
	0-4-0ST	OC	P	1585	1922 New	Scr c/1955
	0-4-0ST	OC	P	2112	1949 New	(1)
	0-4-0ST	OC	P	2136	1953 New	(2)

(a) make unknown, may have been purchased new.

(1) to J.C.Staton & Co Ltd, Tutbury, N Staffs, 3/1958.
(2) Wdn /1964. Scrapped on site by R. Storer, Derby, 1/1966.

WAGON REPAIRS LTD.
BRANSTON WORKS, BURTON ON TRENT.   SH14
Registered 4/1918

This wagon works (SK234221) was located in the triangle of lines where the Midland Railway Leicester line joined the Birmigham to Derby line. The works were erected before 1918 for the Birmingham Railway Carriage & Wagon Co Ltd and sidings connected with the Midland Railway near Branston Sidings No.1 Signal Box. Works closed 10/1968.

Gauge: 4ft 8½in

MT 23	4wPM	BgE	2071	1931 New	(1)
(MT 77) L8	4wDM	RH	393303	1956 New	(2)

(1)   to Gloucester Works, c/1948.
(2)   to Wellingborough Works, Northants, 2/1968.

WAR DEPARTMENT
CANNOCK CHASE MILITARY RAILWAY.   SA6

This railway was constructed during 1915 to serve the Brocton and Rugeley Military Camps located on Cannock Chase. One line was constructed during the spring of 1915 from the LNWR Cannock to Rugeley line near West Cannock No.5 colliery across the Chase to the Rugeley camp. Between January and April a second railway was made from the LNWR Trent Valley line at Milford to the Brocton camp (SJ979187) and by mid 1915 the lines had been joined. In addition to army and prisoner of war camps this railway system served Central Stores Depots (CSD 4 and 375) at Brocton camp. The locomotive shed was also located at Brocton camp. After the war the camps and railway were dismantled and locomotives disposed of.

Ref: 'A Town for Four Winters' by C.J and G.P.Williams

Gauge: 4ft 8½in

	MESSENGER	0-6-0ST	IC	MW	166	1875	(a)	(1)
	BLACKCOCK	0-4-2ST	IC	BP	1140	1871	(a)	(1)
4094		0-6-0T	IC	HC	319	1889	(b)	(5)
4096		0-6-0ST	IC	HC	333	1890	(c)	(6)
WD 85	PYRAMUS	0-6-2T	OC	HL	2879	1911	(d)	(2)
WD 86	MONMOUTH	0-6-0ST	IC	HE	397	1886	(d)	(4)
WD 92	AVONSIDE	0-6-0ST	OC	AE	1742	1916 New	(3)
		0-6-0ST	IC	MW	812	1881	(e)	(7)

The following locomotives are identified in spares orders to Trollope & Sons & Colls & Sons, Brocton Camp, Stafford, and are presumed to have been used in the operation of the line:

No.7	GRASSHOLME	0-6-0ST	IC	MW	1513	1901	(f)	(8)
	UXBRIDGE	0-6-0ST	IC	HE	761	1902	(g)	(9)

(a)   ex West Cannock Colliery Co Ltd, loan, c/1915.
(b)   ex Price and Wills, GCR Immingham Dock contract, Lincs.
(c)   ex Balfour, Beatty Ltd, Ripon Camp contract, Yorks (NR).
(d)   ex Kinmel Camp Railway, near Rhyl, Flints, c/1915.
(e)   ex ? ; orig Lucas & Aird, Hull contract.
(f)   ex Thomas Summerson & Sons Ltd, Albert Hill Foundry, Darlington, Co Durham.
(g)   ex John Wilson & Sons, contrs, Birmingham, after 11/1913, by 7/1915.

(1)   returned to West Cannock Colliery Co Ltd, c/1918.
(2)   for sale 4/1921; to Frank Edmunds, dealer, Stoke on Trent, 8/1921 and resold to Mersey Docks & Harbours Board.
(3)   to Chilwell Depot, Nottinghamshire, c/1922.
(4)   For sale by Ministry of Munitions Disposals Board, 2/6/1919, lying at Crewe Works LNWR; to WD Catterick Camp, Yorkshire (NR) (by 1/4/1924) via J F Wake, dealer, Darlington, /1919. Later Carlton Collieries Association, Barnsley, Yorks (WR).
(5)   to West Cannock Colliery Co Lt, after 11/1920.
(6)   for sale at CSD 4 (Cannock Chase), 9/1921; s/s
(7)   to Royal Arsenal, Woolwich, London.
(8)   May have remained property of Summerson and either returned to him or sold direct to Vivian & Sons Ltd, Mynydd Newydd Colliery, Swansea.
(9)   returned to John Wilson & Sons by 12/1916.

WAR DEPARTMENT
FEATHERSTONE FACTORY, Coven, Nr Wolverhampton   SB3
Ministry of Aviation
Royal Ordnance Factory, Paradise Factory until 1/1960

Factory (SJ927058) built in 1941 beside the LMS Wolverhampton to Stafford line at Featherstone by Paulings Ltd (which see). Extensive sidings were laid within the site. Originally known as ROF Paradise, depot later became known as WD Featherstone and closed c/1963. All track was lifted and site became a HM Government Open Prison.

Gauge: 4ft 8½in

ROF 17 No.1		0-4-0ST	OC	P	2017	1941	New	(1)
ROF 17 No.2		0-4-0ST	OC	P	2019	1941	New	(1)
		0-4-0DM		JF	22000	1937	(a)	(2)
ROF 15 No.2		0-4-0DM		JF	22982	1942	(b)	(3)
ROF 15 No.3	WD8304	0-4-0DM		JF	22989	1942	(b)	(4)
	WD859	0-4-0DM		AB	357	1941	(c)	(5)

(a)   ex Bescot Factory, Staffordshire (see West Midlands Handbook).
(b)   ex Bramshall Factory, Uttoxeter, N Staffs, 8/1948.
(c)   ex WD Bicester, Oxon, 14/6/1961.

(1)   to W H Arnott, Young & Co Ltd, for scrap, 10/1963.
(2)   to Elstow Factory, Bedford, 12/1949.
(3)   to WD Bicester, Oxon, 8/1961.
(4)   to W H Arnott, Young & Co (Shipbreakers) Ltd, Dalmuir, Dunbartonshire, c/1964.
(5)   to WD Ruddington, Notts, 1/4/1964.

WEST CANNOCK COLLIERY CO LTD
Registered 19/11/1869

HEDNESFORD COLLIERIES or
WEST CANNOCK No.1 COLLIERY.	SC12
WEST CANNOCK No.2 (First site) COLLIERY.	SC13
WEST CANNOCK No.2 (orig. No.4) COLLIERY.	SC14
WEST CANNOCK No.3 COLLIERY.	SC15

The West Cannock Colliery was promoted by William Molyneaux and was commenced in 1869 on land leased from the Marquis of Anglesey. Sinking of the No.1 mine (SJ994129) was completed by 1871 and a railway was built to link the pit with the LNWR Walsall to Rugeley line south of Hednesford Station. The line crossed the Cannock to Hednesford Road on the level (later replaced by a bridge under the road) before joining the exchange sidings with the LNWR. Here locomotive sheds and workshops (SJ998123) were built. No.3 (SJ995123) and No.4 (SJ994131) pits were later sunk, respectively, south east of and adjacent to the No.1 mine and served by the same railway. The No.4 pit, which by Vesting Day had become known as No.2 Pit, comprised only one shaft. Often this mine was grouped together with the No.1 colliery as the No.1 plant. A short lived No.2 pit (SJ997117), closed by 1894, was also sunk on the opposite side of the LNWR line and a separate siding was laid to serve this. There were several narrow gauge tramways connected with these collieries. Main and tail haulage was employed on a one and a quarter mile long pit tub tramway laid from the No.1 pit to a landsale wharf at Huntington (SJ975116). Another rope worked tramway ran from No.1 past No.3 under the Cannock Road and the LNWR and through a cutting to a wharf beside the Cannock Extension Canal (SJ996111). Underground endless rope haulage was used in the workings. No.3 pit was closed for a time but reopened about 1890. About this time the standard gauge track was taken under the Cannock Road instead of across it. Further rebuilding of the mine was carried out between 1920 and 1923. A considerable wagon fleet, which at one time numbered 1000 wagons, was operated by the company and these were maintained in a wagon shop beside the locomotive shed. Mines passed to NCB, WM Division, No.2 Area, 1/1/1947 as West Cannock 1-3 Collieries.

BRINDLEY HEATH COLLIERY or
WEST CANNOCK NO.5 COLLIERY   SC16

Sinking of the No.5 pit (SK007141) began in 1914 and sidings were laid to connect with the LNWR Cannock to Rugeley line north of Hednesford Station. Work was halted in 1915 with the onset of war and the railway and some of the locomotives were requisitioned by the WD to help build and run the Cannock Chase Railway(which see). Work on the mine recommenced in 1919 and new sidings and screening plant were laid down. Electric haulage engines were installed underground, but unfortunately the shafts had been sunk in the wrong place and an underground drift had to be made to reach the coal measures. By 1923 the mine was in full production. Locomotives which shunted the mine were based at a shed located at SK 007142. Mines passed to NCB, WM Division, No.2 Area, 1/1/1947 as West Cannock No.5 Colliery.

Gauge: 4ft 8½in

No.3

BLACKCOCK	0-4-2ST	IC	BP	1140	1871	New	(1)
	0-4-0ST	OC				(a)	s/s
	0-6-0ST	OC	P	879	1901	New	(1)
MESSENGER	0-6-0ST	IC	MW	166	1865	(b)	(2)
STAFFORD	0-6-0T	IC	HC	319	1889	(c)	(3)
TOPHAM	0-6-0ST	OC	WB	2193	1922	New	(3)

(a) ex ?
(b) ex Cannock & Rugeley Colliery Co Ltd, c/1914.
(c) ex WD Cannock Chase Railway, after 11/1920.

(1) to NCB, 1/1/1947, Brindley Heath Colliery.
(2) loaned to WD Cannock Chase Railway, c/1915. Retn c/1918 and Scr /1922.
(3) to NCB, 1/1/1947, Hednesford Colliery.

WORTHINGTON & CO LTD
BURTON ON TRENT BREWERIES.
Registered 1/1889

SH15

The original brewery (SK251230), founded by William Worthington in 1760, stood in High Street and during the 1860's this brewery was connected to the Midland Railway Hay Branch. All traffic was handled by the Midland Railway and it was not until 18th November 1872 that Worthington purchased a locomotive to deliver its wagons across High Street onto the Midland Railway Hay Branch. About 1881 another rail connection was made into the premises when the New Street branch was laid. This short branch served several small breweries before connecting with the Bond End branch. In 1914 Worthington absorbed the Burton Brewery Co Ltd (which see). Up to 1924 a small fleet of four locomotives was maintained to work alongside about 30 horses over 5 miles of internal track. Locomotives were based at a shed (SK249230) within the brewery and later also used Salt's old shed (SK254236). Some of the petrol/diesel locomotives were outstationed, one at Crown Maltings, Anglesey Rd (SK238226) (**SH35**) which was reached by the Bond End branch and another at Shobnall. The business was merged with Bass, Ratcliff & Gretton Ltd and from 27/5/1960 locomotive stocks were pooled with those of Bass. Renumbering of Worthington locomotives into the Bass series began in 1960 and all remaining steam locomotives were transferred to the Bass loco shed in 1961. By 1961 Worthington and Bass locomotives worked indiscriminately over both systems. For later history see under Bass.

Gauge: 4ft 8½in

1	0-4-0ST	IC	Crewe	1473	1872	(a)	(1)
(2) 3	0-4-0ST	OC	HC	262	1883	New	(2)
	0-4-0ST	OC	HC	452	1896	New	
	Reb	HC			1936		(3)
4	0-4-0ST	OC	HC	602	1901	New	
	Reb	HC 1911, 1925 & 1952					(4)
2	0-4-0ST	OC	HC	690	1904	New	
	Reb	HC	1934 and 1954				(4)
5	0-4-0ST	OC	MW	1427	1899	(b)	
	Reb	MW			1916		(5)

South Staffordshire Handbook. Page 82

5	0-4-0ST	OC	WB	2108	1923	New	
	Reb		WB	1933 and	1955		(4)
6	0-4-0ST	OC	HC	1417	1920	New	
	Reb		HC		1947		(4)
7	4wPM		KC		1924	New	
	Reb 4wDM				1953		(4)
8	4wPM		KC		1924	New	
	Reb 4wDM						(4)
9	4wPM		KC		1925	New	
	Reb 4wDM						(4)
10	4wPM		KC		1926	New	
	Reb 4wDM						(4)
	4wPE		Moyse	86 ?	1926	New	s/s
11	4wPM		FH	1612	1929	New	
	Reb 4wDM				1950		(4)
1	0-4-0ST	OC	HC	272	1884		
	Reb		HC		1923	(c)	Scr /1954
12	4wPM		FH	1846	1934	New	
	Reb 4wDM				1952		(4)
1	0-4-0ST	OC	WB	2815	1945	New	(4)

(a) ex LNWR, 1205, to work 19/11/1872.
(b) ex Burton Brewery Co Ltd, /1914.
(c) ex Thomas Salt and Co Ltd, Burton, /1928.

(1) for sale 10/1883, s/s.
(2) to Tyne Tees Shipping Co Ltd, Middlesborough,, /1929.
(3) to Bass Ratcliff and Gretton Ltd, 4/1954.
(4) to Bass Ratcliff and Gretton Ltd, 27/5/1960.
(5) to Mansfield Standard Sand Co Ltd, Mansfield, Notts, /1926.

SECTION 3

BRITISH COAL CORPORATION (NATIONAL COAL BOARD until 5/3/1987)

Hilton Main
and Holly Bank
COLLERIES
LIMITED

MITRE COALS

Local Offices:

46 QUEEN STREET
WOLVERHAMPTON

COLLIERIES:
HILTON & ESSINGTON

Telephone—21893   Telephone—21893

PRICE LISTS UPON APPLICATION

BRITISH COAL CORPORATION - NATIONAL COAL BOARD

On the nationalisation of the coal industry on 1st January 1947, the assets of the colliery companies were vested in the National Coal Board. Small mines with less than 30 men underground could be excluded and operate under license from the NCB. There were a number of these in this Coalfield. Those companies nationalised had the option to include or exclude ancillary operations such as brickworks; such subsidiary activities in Staffordshire were generally included. The title, National Coal Board remained in use until 1986 when the trading name British Coal was adopted. The legal title became British Coal Corporation from 5th March 1987.

DEEP MINES ORGANISATION.

Collieries absorbed by the NCB on Vesting Day were placed in an organisation comprising eight Divisions, each of which was divided into a number of Areas. Each Area was in turn sub-divided into Groups. However, Groups had little railway significance and reorganisations were frequent, hence these early Groups are disregarded in IRS publications.

The collieries in South Staffordshire fell mainly into Area No.2 of the West Midlands Division, with headquarters at Hednesford and later at Cannock. One colliery, Amington, was included in the No.4 (Warwickshire) Area with Headquarters at Nuneaton. From the 26th March 1967, a major reorganisation replaced Divisions by eighteen 'New style' Areas. Surviving South Staffordshire collieries were joined with those of North Staffs and Shropshire to form Staffordshire Area. On 1st April 1974, Staffordshire Area amalgamated with North Western Area to form Western Area.

The next reorganisation, effective from 1st. January 1990, drastically reduced Area staff and much day to day control reverted to collieries. Areas became known as Groups and Western Area's South Staffordshire collieries passed to Central Group with Headquarters at Coleorton, Leicestershire. As contraction of the mining industry continued, a further change has been an amalgamation of Groups. From 1st October 1991, North Western, Central and South Wales Groups were combined to form a new Midlands and Wales Group. Headquarters are at Beaumont House, Coleorton, Leicestershire.

WORKSHOPS ORGANISATION.

Area Central Workshops were part of the deep mines organisation until June 1967, after which they passed to the direct control of National Headquarters.

COKING, TAR AND MANUFACTURED FUEL PLANTS.

There have been no plants which fall into this category in South Staffordshire.

BRICKWORKS.

Administered by the Divisions of the Deep Mines organisation until 1st January 1962 when they were combined nationally to form the Brickworks Executive. This operated through subsidiary companies, the one covering Staffordshire being known as the Midland Brick Co.Ltd. By 1955, in Staffordshire only the works at Hednesford and Hilton Main remained open. To these was added the Hawkins works at Bridgtown which was purchased in the 1960's. The subsidiary companies of the Brickworks Executive were sold to the private sector on 25th November 1973 and the executive was later wound up.

OPENCAST WORKINGS.

Opencast coal sites were taken over by the newly formed Opencast Executive of the National Coal Board from 1st April 1952. Prior to this, they had been administered by the Ministry of Fuel and Power, Directorate of Opencast Coal Production. At that time several sites were in production in the Cannock area of South Staffordshire. They formed part of Region No.9 (later No.6) (South Midlands) of the Opencast Executive. From about 1967 there were no sites in operation. Further rail served sites were operated from the late 1970's, by which time the region had been retitled Central West

Opencast coal has been sold from Disposal Points which may serve several sites and may or may not be rail served. No rail served disposal points are known for the sites operated before 1958. Disposal points are let out to private contractors who often own or hire any locomotives used.

LOCATION LISTINGS

Locations which have made any use of rail track, surface or underground, standard or narrow gauge, are listed below alphabetically, together with brief notes of the rail system and full details of locomotives where appropriate. The following abbreviations are used to save space when showing in which part of the organisation a location was placed at any given period:-

WM2	West Midlands Division, Area No.2.
WM4	West Midlands Division, Area No.4.
STF	Staffordshire Area.
WES	Western Area.
CEN	Central Group.
MWG	Midlands and Wales Group.
HQ	National Headquarters, Workshops Organisation.
OE	Opencast Executive.

AMINGTON COLLIERY, Amington.   SK4
Ex Glascote Colliery Co.Ltd.   (SK241035)

WM4 from 1/1/1947; Merged with **POOLEY HALL** and **ALVECOTE** (Warwickshire) Collieries to form **NORTH WARWICK** Colliery 5/1951 and rail system CLOSED. Shafts retained for manriding and ventilation until the combined mine CLOSED 4/1965.

An NCB branch ran south from BR., 2 miles east of Tamworth (Low Level) Station to the colliery (¾ miles). It continued west to join the railway of Messrs. Gibbs & Canning Ltd. at the site of the closed Glascote Colliery (¾ miles) and thence further west to a wharf on the Coventry canal and a second junction with BR, ½ mile south of Tamworth (High Level) Station (2 miles from colliery).
Note; The section west of Glascote was used by Gibbs & Canning until the railway closed. It is not clear if there was any NCB traffic on this section after vesting day. Locomotives were not used underground.

Gauge: 4ft 8½in

AMINGTON No.2	0-6-0ST	OC	HL	3642	1925	(a)	(1)
AMINGTON No.3	0-6-0ST	OC	WB	2508	1934	(a)	(2)

(a)   ex Glascote Colliery Co Ltd, 1/1/1947.

(1)   to Kingsbury Colliery, Warwickshire, 6/1951.
(2)   to Pooley Hall (North Warwickshire) Colliery, 6/1951.

BAGGERIDGE COLLIERY, Sedgeley.   SJ3
Ex Earl of Dudley's Baggeridge Collieries Ltd.   (SO899928)

WM3 from 1/1/1947; WM2 from 1/1/1962; STF from 26/3/1967; CLOSED 3/1968.

A branch ran north from BR, 1½ miles east of Himley Station, to the colliery (1½ miles). This connected 1 mile south of the colliery, with the Pensnett Railway of the Earl of Dudley's Round Oak Steelworks Ltd. Round Oak locomotives worked all traffic to both BR and the steelworks until 1952 when the NCB introduced its own locomotives for the traffic to BR and the colliery shunting. Round Oak locomotives worked coal to the steelworks, 4 miles fom the colliery until the colliery closed. One underground diesel locomotive was delivered here in 1955 but moved on shortly afterwards.

Gauge: 4ft 8½in

4wDM		RH	321730	1952	New	(1)
0-6-0ST	IC	HE	3776	1952	New	(2)
0-6-0ST	IC	HE	3777	1952	New	(3)
4wDM		RH	338413	1953	(a)	(4)

(a)   ex Mid Cannock Colliery, between 3/1967 and 3/1968.

(1) to Sandwell Park Colliery, West Midlands, /1952, returned by 8/1953. To Mid Cannock Colliery, between 10/1963 and 4/1964.
(2) to BR Wolverhampton Works, repairs after 3/1960, returned /1960. To Hilton Main Colliery, between 8/1967 and 2/1968.
(3) to Florence Colliery, Longton, N Staffs, w/e 30/3/1968.
(4) to Hilton Main Colliery, c2/1969.

Underground locomotives
Though rope haulage was used in the mine, one locomotive was delivered to the pit in 1955. It is not known if it actually went underground.

Gauge: 2ft 0in

                    0-4-0DMF   RH   388770   1955 New   (1)

(1) to Binley Colliery, West Midlands, 6/1955.

BRERETON COLLIERY, Rugeley.   SA11
Ex Brereton Collieries Ltd.   (SK044152)

WM2 from 1/1/1947; CLOSED 7/1960.

An NCB branch ran south from BR, south west of Rugeley Town Station to the colliery (1½ miles). The locoshed was to the east of this line, ¼ mile north of the colliery (SK046154). A dirt tip east of the colliery at SK048153 was served by rail. Locomotives were not used underground.

Gauge: 4ft 8½in

	VANGUARD	0-4-0ST	OC	P	1491	1917	(a)	(1)
No.3		0-4-0ST	OC	AB	1365	1914	(a)	(2)
B2C		0-4-0ST	OC	Butt		1889	(a)	Scr 9/1952
	BIRCH	2-4-0T	OC	Rawnsley		1888	(b)	(4)
	FOGGO	0-4-2ST	IC	Chasetown		1946	(c)	(3)

(a) ex Brereton Collieries Ltd, 1/1/1947.
(b) ex Rawnsley, loan, after 2/1948, returned by 11/1949. Ex Rawnsley, again, between 11/1949 and 3/1950.
(c) ex Coppice Colliery, between 4/1954 and 8/1954.

(1) to Rawnsley Loco Shed, 8/1959.
(2) to Rawnsley Loco Shed, between 5/1960 and 7/1961.
(3) Scr by W.H.Arnott Young & Co Ltd, 1/1961.
(4) Scr between 2/1956 and 8/1957.

CANNOCK CENTRAL STORES, Chase Terrace.　　　　　　　　　SE16
(SK035093?)

Opened by WM2 in 1950's; STF from 26/3/1967; HQ from 6/1967; CLOSED

Situated adjacent to **CANNOCK CENTRAL WORKSHOPS**. No rail connection. A new surface narrow gauge locomotive was delivered here pending allocation to a colliery.

Gauge: 2ft 6in

4wDM	RH	7002/0867/3	1967	New	(1)

(1)　to Hilton Main Colliery,　/1967

CANNOCK CENTRAL WORKSHOPS, Chase Terrace.　　　　　　　SE16
Ex Cannock Chase Colliery Co.Ltd.　　　　　　　　　　　　　　(SK035092)

WM2 from 1/1/1947; STF from 26/3/1967; HQ from 6/1967; CLOSED

Connected to the **CANNOCK CHASE COLLIERIES** rail system (which see) until that CLOSED c4/1962. Locomotive repairs were carried out until c/1965.

Gauge: 4ft 8½in

No.	Name	Type	Cyl	Builder	Works	Date	Ref	No.
No.3		0-6-0ST	IC	P	618	1895	(a)	(1)
	LITTLETON No.6	0-6-0ST	IC	RSH	7292	1945	(b)	(2)
No.8		0-6-0ST	IC	RSH	7106	1943	(c)	(3)
1	MARQUIS	0-6-0ST	IC	Lill		1867	(d)	(4)
No.3	HANBURY	0-6-0ST	IC	P	567#	1894	(e)	(5)
2		0-6-0T	IC	K	5358	1921	(f)	(6)
	NUTTALL	0-6-0ST	OC	HE	1685	1931	(g)	(7)
No.2		0-6-0ST	IC	HE	3772	1952	(h)	(8)
No.3		0-6-0ST	IC	HE	3789	1953	(i)	(9)
7	WIMBLEBURY	0-6-0ST	IC	HE	3839	1956	(j)	(10)
8		0-6-0ST	IC	HE	3807	1953	(k)	(11)
	GRIFFIN	0-6-0ST	IC	K	5036	1913	(l)	(12)
No.6	ADJUTANT	0-6-0ST	OC	MW	1913	1917	(m)	(13)
	THE COLONEL	0-6-0ST	IC	HC	1073	1914	(n)	(14)
	HAWKINS	0-6-0ST	OC	P	809	1900	(o)	(15)
	HOLLYBANK No.3	0-6-0ST	IC	HE	1451	1924	(p)	(16)
	LITTLETON No.1	0-6-0ST	IC	MW	1515	1901	(q)	(17)
	CONDUIT No.3	0-6-0ST	IC	MW	1180*	1890	(r)	(18)
	CONDUIT No.1	0-6-0ST	IC	MW	244	1867	(s)	(19)
	AYNHO	0-6-0ST	IC	MW	1722	1908	(t)	(20)
	TOPHAM	0-6-0ST	OC	WB	2193	1922	(u)	(21)
	STAFFORD	0-6-0T	IC	HC	319	1889	(v)	(22)
	CONDUIT No.4	0-6-0ST	IC	MW	1326	1896	(w)	(23)
	LORD KITCHENER	0-6-0ST	IC	K	5158	1915	(x)	(24)
No.1		4wDM		RH	338413	1953	(y)	(25)

\#　Incorporated parts of P 618 from　/1963
*　Incorporated parts of, and plates from, MW 1326 from　/1957

(a)   ex Grove Colliery, between 3/1963 and 4/1963.
(b)   ex Littleton Colliery, between 12/1957 and 4/1958.
(c)   ex Chasetown Loco Shed, between 4/1959 and 6/1959.
(d)   ex Grove Colliery, between 3/1963 and 4/1963.
(e)   ex Walsall Wood Colliery, West Midlands, between 10/1951 and 5/1953.
(f)   ex Coppice Colliery, between 5/1953 and 1/1954.
(g)   ex Walsall Wood Colliery, 8/1956
(h)   ex Chasetown Loco Shed, between 8/1957 and 4/1958.
(i)   ex Chasetown Loco Shed, between 2/1962 and 5/1962.
(j)   ex Rawnsley Loco Shed, 3/1963.
(k)   ex Rawnsley Loco Shed, between 5/1958 and 12/1958.
(l)   ex Walsall Wood Colliery, between 5/1955 and 9/1955.
(m)   ex Rawnsley Loco Shed, between 11/1955 and 8/1956.
(n)   ex Grove Colliery, between 5/1960 and 6/1961.
(o)   ex Hawkins Colliery, between 4/1960 and 4/1961.
(p)   ex Littleton Colliery, between 3/1959 and 6/1959.
(q)   ex Littleton Colliery, between 2/1954 and 9/1955.
(r)   ex West Cannock 1-3 Collieries, between 10/1955 and 8/1956.
(s)   ex Conduit Colliery, between 1/1950 and 5/1950
(t)   ex Walsall Wood Colliery, between 10/1951 and 12/1951.
(u)   ex West Cannock No.5 Colliery, between 5/1960 and 6/1961.
(v)   ex West Cannock 1-3 Collieries, between 3/1957 and 7/1957.
(w)   ex Coppice Colliery, between 10/1050 and 8/1951.
(x)   ex Walsall Wood Colliery, between 6/1951 and 10/1951.
(y)   ex East Cannock Colliery, 20/8/1957.

(1)   parts incorporated in P 567 and the remains scr on site, c5/1963.
(2)   to Littleton Colliery, between 12/1958 and 3/1959.
(3)   to Chasetown Loco Shed, between 6/1959 and 10/1959. Ex Chasetown, between 2/1962 and 5/1962, to Littleton Colliery, 19/4/1963.
(4)   scr on site by L.Wallace, of Cannock, 5/1964.
(5)   to Coppice Colliery, between 4/1954 and 8/1954. Ex Coppice Colliery, between 12/1958 and 4/1959, to Littleton Colliery, between 3/1960 and 12/1960. Ex Coppice Colliery, 2/5/1963, to Hilton Main Colliery, 3/4/1964.
(6)   to Coppice Colliery, between 3/1955 and 5/1955. Ex Coppice Colliery, 13/7/1963; Scr on site by L.Wallace, of Cannock, 7/1964.
(7)   to West Cannock 1-3 Colliery, 31/5/1957. Ex Chasetown Loco Shed, between 2/1962 and 5/1962, to Hollybank Shed 26/10/1962.
(8)   to Chasetown Loco Shed between 10/1958 and 12/1958. Ex Chasetown, between 2/1962 and 5/1962, to Hollybank Shed, between 5/1962 and 7/1962.
(9)   to Rawnsley Loco Shed, 4/1963.
(10)   to Rawnsley Loco Shed, /1964.
(11)   to Rawnsley Loco Shed, between 12/1958 and 6/1959.
(12)   to Walsall Wood Colliery, 28/3/1956.
(13)   to Rawnsley Loco Shed, between 11/1956 and 8/1957. Ex Rawnsley, between 9/1960 and 6/1961. Scr on Site 3/1962.
(14)   to Coppice Colliery, 11/7/1963.
(15)   Scr on site by W.H.Arnott Young and Co Ltd, 15/5/1961.
(16)   to Littleton Colliery, between 5/1960 and 6/1961.
(17)   to Littleton Colliery between 11/1955 and 8/1957.
(18)   to West Cannock No.5 Colliery, between 7/1957 and 11/1957.
(19)   Scr /1952, after 10/1951
(20)   Scr /1955, after 3/1955
(21)   to West Cannock No.5 Colliery, 7/9/1962.

(22) to West Cannock 1-3 Collieries, between 12/1958 and 6/1959.
(23) parts incorporated into MW 1180, remainder scr on site /1957.
(24) to Walsall Wood Colliery, 11/1951. Ex Walsall Wood Colliery, between 8/1958 and 4/1959. To Walsall Wood Colliery, 4/1963.
(25) to Mid Cannock Colliery, 5/1958.

Gauge: 2ft 6in

	4wDM	RH		1944	(a)	(1)
	4wDM	RH	466587	1961	(c)	(2)
	4wDM	RH	441947	1959	(b)	(3)
No.2	4wBEF	EE			(d)	(4)

(a) ex Wyrley No.3 Colliery, /1962, by 5/1962.
(b) ex Lea Hall Colliery, 27/12/1961.
(c) ex Lea Hall Colliery, 4/1963
(d) unidentified underground loco noted under repair, 10/1969

(1) to Wyrley No.3 Colliery, /1962.
(2) to Lea Hall Colliery, c/1964
(3) to Lea Hall Colliery, 22/2/1962.
(4) to ? after 10/1969

CANNOCK CHASE COLLIERIES, Comprising:- SE3-9
CANNOCK CHASE No.3 COLLIERY, Chase Terrace. (SE3) (SK035090)
CANNOCK CHASE No.7 COLLIERY, Chase Terrace. (SE7) (SK037111)
CANNOCK CHASE No.8 COLLIERY, Heath Hayes. (SE8) (SK022105)
CANNOCK CHASE No.9 COLLIERY, Hednesford. (SE9) (SK008114)
Ex Cannock Chase Colliery Co.Ltd.

WM2 from 1/1/1947; **No.7 Colliery** was only a service shaft for **No.8 Colliery** by Vesting Day. **No.9 Colliery** merged with **No.3 Colliery** and No.9 surface CLOSED 1951; No.3 and No.8 Collieries were managed as one unit. After No.3 CLOSED 1/1959, both surfaces remained until production ceased from No.8 Colliery on 26/1/1962. The washery at No.3 CLOSED on 23/2/1962.

An NCB branch ran north west from BR at Anglesea Sidings (SK 055065) (1 mile west of Hammerwich Station), for 1¼ miles to Chasetown Locoshed (SK040082), thence for a further ½ mile to No.3 Colliery and the washery for the system. The Central Workshops and Stores were located to the east of No.3 Colliery. The line continued north west for a further 1 mile to No.8 Colliery which was also connected to No.3 Washery by an overland 2ft0in gauge rope hauled tubway. A branch ran north east from a point between the locoshed and No.3 Colliery for ½ mile to a landsale yard at Chase Terrace **(SE5)**, continuing north for a further mile to No.7 shaft and an adjacent brickworks and then to a junction with the BR Cannock Chase Railway **(SC2)** south of Rawnsley. Running powers over this and the connecting Littleworth Tramway **(SC1)** enabled No.9 Colliery (4 miles from Chasetown) and a nearby brickworks to be served. Running powers were also held over the BR Littleworth Extension Railway which ran west from No.9 Colliery to BR's Norton Branch (¾ mile). These running powers ceased to be used by Chasetown locos c/1951 after No.9 Colliery closed and the NCB line north of No.7 shaft closed. Note; NCB locomotives from the Rawnsley shed of the Cannock Wood System (which see) exercised similar running powers until c1962. Another NCB branch also diverged between the locoshed and No.3 Colliery. This curved west for 1 mile, partly on a causeway across Norton Pool, to connect with the northern end of the BR Brownhills (Watling Street) branch and with the BR Fiveways branch via a further ½ mile NCB line. The standard gauge line to No.8 Colliery and the line north of Chase Terrace landsale depot had closed by 1960. The remaining rail system closed with No.3 Washery on 23/2/1962, and was lifted except for a few sidings at Central Workshops, which lasted for several more years, and the line which crossed Norton Pool to the sidings at its western end. The latter sidings became part of the preserved Chasewater Light Railway (which see) Locomotives were not used underground at any of the Cannock Chase Collieries

Gauge: 4ft 8½in

	Mc CLEAN	0-4-2ST	IC	BP	28	1856	(a)	(1)
	ALFRED PAGET	0-4-2ST	IC	BP	204	1861	(a)	(2)
	CHAWNER	0-4-2ST	IC	BP	462	1864	(a)	(3)
	ANGLESEY	0-4-2ST	IC	BP	1211	1872	(a)	(4)
No.6		0-6-0ST	IC	SS	2643	1876	(a)	(5)
	GRIFFIN	0-6-0ST	IC	K	5036	1913	(a)	(6)
75070		0-6-0ST	IC	RSH	7106	1943	(a)	(7)
	FOGGO	0-4-2ST	IC	Chasetown		1946	(a)	(8)
		0-4-0ST	OC	AB	2247	1948	New	(9)
2		0-6-0ST	IC	HE	3772	1952	New	(10)
3		0-6-0ST	IC	HE	3789	1953	New	(11)
4		0-4-0ST	IC	HE	3806	1953	(b)	(12)
	NUTTALL	0-6-0ST	OC	HE	1685	1931	(c)	(13)
	TONY	0-6-0ST	OC	HL	3460	1921	(d)	(14)

(a) ex Cannock Chase Colliery Co Ltd.,1/1/1947.
(b) ex Rawnsley Shed /1953.
(c) ex West Cannock Nos.1-3 Collieries, 2/7/1957.
(d) ex Hawkins Colliery between 11/1955 and 10/1958.

(1) to Coppice Colliery after 3/1948, returned by 11/1949. Scr between 11/1955 and 2/1956.
(2) Scr between 10/1951 and 5/1952
(3) Scr between 3/1948 and 11/1949
(4) Scr between 4/1951 and 10/1951
(5) Scr between 10/1951 and 5/1952
(6) to Walsall Wood Colliery between 10/1951 and 5/1953.
(7) to Cannock Central Workshops between 4/1959 and 6/1959, returned between 6/1959 and 10/1959. To Cannock Central Workshops between 2/1962 and 5/1962.
(8) to Coppice Colliery after 3/1948, returned by 11/1949. To Coppice between 1/1954 and 4/1954.
(9) to Coppice Colliery between 3/1959 and 6/1959.
(10) to Cannock Chase Workshops between 8/1957 and 4/1958, returned. between 10/1958 and 12/1958. To Cannock Chase Workshops for storage between 2/1962 and 5/1962.
(11) to Cannock Chase Workshops for storage between 2/1962 and 5/1962.
(12) to West Cannock No.5 Colliery between 9/1961 and 1/1962.
(13) to West Cannock No.5 Colliery, 8/1957; ex West Cannock No.5 Colliery, between 11/1957 and 3/1958; to Cannock Chase Workshops for storage, between 2/1962 and 5/1962.
(14) to Walsall Wood Colliery 3/2/1959.

South Staffordshire Handbook.

CANNOCK & LEACROFT COLLIERY, Cannock.   **SD30**
Ex Cannock & Leacroft Colliery Co.Ltd.   (SJ997097)

WM2 from 1/1/1947; Merged with **MID CANNOCK COLLIERY** 9/1954.

Sidings ran south from the BR Norton branch 1 mile south of its northern junction at Cannock, to the colliery. A narrow gauge rope worked line ran south west from the colliery to a wharf on a basin off the Cannock Extension Canal (SK 994088) (½ mile). The sidings were worked by BR locomotives and electric capstans. Locomotives were not used underground until after the merger with Mid Cannock (which see). Coal winding and rail traffic ceased on this merger

CANNOCK OLD COPPICE COLLIERY - see HAWKINS

CANNOCK & RUGELEY COLLIERIES, Comprising:-   **SC3 - 5**
    CANNOCK WOOD COLLIERY, Rawnsley.   **(SC3)**   (SK035126)
    VALLEY COLLIERY & TRAINING CENTRE, Hednesford. **(SC4)**   (SK008127)
    WIMBLEBURY COLLIERY, Littleworth.   **(SC5)**   (SK014117)
Ex Cannock & Rugeley Colliery Co Ltd.

WM2 from 1/1/1947; **VALLEY COLLIERY** was only a service shaft for **WIMBLEBURY COLLIERY** by vesting day. **WIMBLEBURY** merged with **WEST CANNOCK No.5 COLLIERY** from 12/1962 and its surface CLOSED. **VALLEY** became a TRAINING CENTRE which passed to STF from 26/3/1967; WES from 1/4/1974; CLOSED by 1988; **CANNOCK WOOD COLLIERY** passed to STF from 26/3/1967; CLOSED 8/6/1973.

The BR owned Cannock Chase Railway **(SC2)** ran east from a junction north of Hednesford Station and turned south east to New Hayes beyond the NCB Rawnsley locoshed (SK 025124) and workshops **(SC19)** (1½ miles). This line was worked exclusively by NCB locos from Rawnsley. At New Hayes there was a connection with the NCB Cannock Chase Colliery Railway (which see). South of Rawnsley the BR line reversed to run west to Wimblebury Colliery (¾ mile) to make an end on connection with the British Waterways owned Littleworth Tramway **(SC1)** which enabled NCB trains from Cannock Wood & Wimblebury to reach a wharf (SJ 997113) on the Cannock Extension Canal (2½ miles from Rawnsley). The lines west of New Hayes were also used by Cannock Chase Colliery trains to serve their No.9 Colliery and a connection to the BR Norton branch until c1951. Cannock Wood Colliery was reached by an NCB line which ran north east from the Cannock Chase Railway at Rawnsley and continued beyond the colliery to rejoin it at New Hayes. This extension was disused by 1960. Valley, which was used only for men and materials, had an NCB line which ran south east from the BR exchange sidings at Hednesford to the colliery (½ mile). This line was closed by 1960. The traffic from Cannock Wood to the canal was carried in wooden containers mounted three to a wagon frame and brakevans were used on these trains. They ceased to run at about the same time that Wimblebury closed (12/1962) and by 1964 all lines south and west of Rawnsley had closed. A new locoshed was opened at Cannock Wood (SK035126) in 10/1964 and that at Rawnsley closed 2/1965. A mineworker's passenger service using a variety of ex main line coaches was operated between Hednesford and Cannock Wood Colliery but had ceased before 1965. This remaining part of the railway closed with Cannock Wood Colliery. As part of a major reconstruction scheme at Cannock Wood, locomotives were installed underground for manriding, minerals and material haulage and in the new surface stockyard, from 1960 and 1959 respectively. Locomotives were not used underground at Wimblebury or Valley Collieries but during Valley's period as a training centre, locomotives were here on two occasions for instructional purposes.

Rawnsley Shed   SC19

Gauge: 4ft 8½in

1	MARQUIS	0-6-0ST	IC	Lill		1867	(a)	(1)
2	ANGLESEY	0-6-0ST	IC	Lill		1868	(a)	(2)
3	PROGRESS	0-6-0ST	IC	P	786	1899	(a)	Scr 12/1964.
4	RAWNSLEY	0-6-0ST	IC	Lill		1872	(a)	(3)
5	BEAUDESERT	0-6-0ST	OC	FW	266	1875	(a)	(4)
6	ADJUTANT	0-6-0ST	OC	MW	1913	1917	(h)	(5)
7	BIRCH	2-4-0T	OC	Rawnsley		1888	(a)	(6)
8	HARRISON	2-4-0T	OC	YE	185	1875		
	Reb to	0-6-0T	OC	Rawnsley		1916	(a)	(7)
9	CANNOCK WOOD	0-6-0T	IC	Bton		1877	(a)	(8)
4		0-6-0ST	IC	HE	3806	1953	New	(9)
8		0-6-0ST	IC	HE	3807	1953	New	(10)
	WIMBLEBURY	0-6-0ST	IC	HE	3839	1956	New	(16)
No.3		0-6-0ST	IC	HE	3789	1953	(b)	(15)
	VANGUARD	0-4-0ST	OC	P	1491	1917	(c)	(11)
		0-4-0ST	OC	AB	1365	1914	(d)	(3)
	STAFFORD	0-6-0T	IC	HC	319	1889	(e)	(12)
	NUTTALL	0-6-0ST	OC	HE	1685	1931	(f)	(13)
	CONDUIT No.3	0-6-0ST	IC	MW	1180#	1890	(g)	(14)

    #   Incorporated parts of, and plates from, MW 1326, from /1957.

(a) ex Cannock & Rugeley Colliery Co Ltd., 1/1/1947.
(b) ex Cannock Central Workshops, 4/1963.
(c) ex Brereton Colliery, 8/1959.
(d) ex Brereton Colliery, between 5/1960 and 7/1961.
(e) ex West Cannock No.5 Colliery, 5/1963.
(f) ex Hollybank Loco Shed, between 10/1962 and 3/1963.
(g) ex Grove Colliery, 7/1963.
(h) ex Littleton Colliery, /1947, where it was on loan from Cannock & Rugeley Colliery Co Ltd at 1/1/1947

(1) to Grove Colliery, 15/1/1962.
(2) to West Cannock No.5 Colliery between 8/1956 and 4/1957, returned between 4/1957 and 7/1957. Scr by Cashmore at Cannock Wood Colliery, 3/1962.
(3) Scr by Cashmore at Cannock Wood Colliery, 3/1962.
(4) to Lea Hall Colliery between 9/1960 and 7/1961, returned Rawnsley between 7/1961 and 5/1962. Wdn /1963. Scr by T.Hill, Chasetown, 6/1964.
(5) to Cannock Central Workshops between 11/1955 and 8/1956; returned between 11/1956 and 8/1957; to Cannock Central Workshops, between 9/1960 and 6/1961.
(6) to Brereton Colliery, loan, after 2/1948. Returned by 11/1949. To Brereton again between 11/1949 and 3/1950.
(7) Scr by Arnott Young and Co Ltd, between 4/1954 and 5/1955
(8) to RPS Hednesford, 12/1963.
(9) to Cannock Chase Loco Shed, /1953.
(10) to Cannock Central Workshops between 5/1958 and 12/1958; returned by 6/1959; to Cannock Wood Colliery, between 10/1964 and 2/1965.
(11) to Hamstead Colliery, West Midlands, 11/1961.
(12) scr by T.Hill, Chasetown, 12/1964.
(13) to Littleton Colliery, 6/1963.
(14) reb with parts of MW 1326, c/1957. Scr 12/1964.

South Staffordshire Handbook. Page 95

(15) to Cannock Wood Colliery, between 10/1964 and 2/1965.
(16) to Cannock Central Workshops, 3/1963; returned /1964; to Cannock Wood Colliery between 10/1964 and 2/1965.

Cannock Wood Colliery   SC3

Gauge: 4ft 8½in

8		0-6-0ST	IC	HE	3807	1953	(a)	Scr c12/1967
No.3		0-6-0ST	IC	HE	3789	1953	(a)	(2)
	WIMBLEBURY	0-6-0ST	IC	HE	3839	1956	(a)	(3)
		0-6-0DH		EEV	D1120	1966	New	(4)
	TOPHAM	0-6-0ST	OC	WB	2193	1922	(b)	(5)
	HEM HEATH No.1	0-6-0ST	OC	WB	3077	1955	(c)	(6)
No.8		0-6-0ST	IC	HE	3776	1952	(d)	(7)
9	CANNOCK WOOD	0-6-0T	IC	Bton		1877	(e)	(8)

(a) ex Rawnsley Shed, between 10/1964 and 2/1965.
(b) ex West Cannock Colliery, 2/1970.
(c) ex Silverdale Colliery, N Staffs, 27/5/1970.
(d) ex Granville Colliery, Shropshire, between 5/1970 and 7/1970.
(e) ex RPS Hednesford Depot, 25/6/70, for loading only.

(1) Scr between 8/1967 and 6/1968
(2) to Granville Colliery, Shropshire, 6/1967.
(3) to Foxfield Railway Preservation Society, Dilhorne, N Staffs, 9/1973.
(4) to EEV 3/1970, returned 8/1970. To Littleton Colliery, 11/10/1973.
(5) to West Cannock No.5 Colliery, 9/1970.
(6) to Norton Colliery, N Staffs, 10/1970.
(7) to West Cannock No.5 Colliery, 1/1971.
(8) to RPS Chasewater Depot, 27/6/1970.

Stockyard Railway

During the reconstruction of the surface plant in the late 1950's a stockyard was constructed on the site of one of the old screens and a narrow gauge railway laid to connect it with the two pitheads still in use.

Gauge: 2ft 6in

	4wDM		RH	452293	1959	New	(1)
	4wDM		RH	466587	1961	New	(2)

(1) to Littleton Colliery, between 12/1970 and 10/1971.
(2) to Lea Hall Colliery by 4/1962. ex Littleton Colliery, between 10/1969 and 10/1971. To Walkden Central Workshops, Gtr Manchester, c/1974.

Underground Haulage

At Cannock Wood mechanical rope haulage was used to reach most of the workings. Horses were also employed underground. It was not until 1960 after the mine was modernised when underground locomotives were employed in the workings.

Gauge: 2ft 6in

4wBEF	EE	2844	1960		
	RSHN	8131	1960	New	(1)
4wBEF	EE	2845	1960		
	RSHN	8132	1960	New	(2)
4wBEF	EE	2846	1960		
	RSHN	8133	1960	New	(3)
4wBEF	EE	2847	1960		
	RSHN	8134	1960	New	(4)
4wBEF	EE	2860	1960		
	RSHD	8203	1960	New	(5)
4wBEF	GB	6090	1963	New	(6)

(1) to Littleton Colliery, 11/1972.
(2) to West Cannock No.5 Colliery, 11/1973.
(3) to Lea Hall Colliery, 8/1960.
(4) to West Cannock No.5 Colliery, 7/1973.
(5) to Littleton Colliery, 28/6/1973.
(6) to Lea Hall Colliery, 1/1974.

Valley Training Centre. SC4

The Valley Colliery (SK008127) was acquired by the NCB 1/1/1947 from the Cannock & Rugeley Colliery Co Ltd. By this time it had ceased drawing its own coal and the seams were worked from Wimblebury. Part of the mine was converted into a NCB Training Centre. Locomotives have been based here for instruction purposes.

Gauge: 2ft 6in

4wBEF	CE	5074	1966	(a)	(1)

(a) ex West Cannock No.5 Colliery, by /1969.

(1) to Lea Hall Colliery, 10/7/1969.

Becorit Road Rail System.

Gauge: 200mm

1ad DMF	BGB	25/2/216	1971	(a)	(1)

(a) ex Littleton Colliery, /1971.

(1) to Kemball Training Centre, Trentham, N Staffs.

CANNOCK WOOD COLLIERY - see CANNOCK & RUGELEY COLLIERIES.

CONDUIT COLLIERY, Norton Canes. SE7
Ex Littleton Collieries Ltd. (SK020081)

WM2 from 1/1/1947; CLOSED 8/1949

Sidings on the west side of the BR Fiveways branch north of its junction with the Norton Canes branch served the colliery and a wharf on an arm of the Cannock Extension Canal. Locomotives were not used underground.

Gauge: 4ft 8½in

	CONDUIT No.1	0-6-0ST	IC	MW	244	1867	(a)	(1)
	CONDUIT No.4	0-6-0ST	IC	MW	1326	1896	(a)	(2)
	AYNHO	0-6-0ST	IC	MW	1722	1908	(a)	(3)

(a)   ex Littleton Collieries Ltd, 1/1/1947.

(1)   to Cannock Central Workshops, between 1/1950 and 5/1950.
(2)   to Coppice Colliery, 10/1949.
(3)   to Walsall Wood Colliery, between 2/1950 and 5/1950.

COPPICE COLLIERY, Norton Canes. SE11
Ex Coppice Colliery Co.Ltd. (SK014096)

WM2 from 1/1/1947; CLOSED 4/1964

Sidings at the north end of the BR Fiveways branch served the colliery. Running powers were used over this branch to gain access to the NCB line which curved east from opposite Conduit Colliery to BR exchange sidings north of Brownhills (Watling Street) Station, where connection was also made with the Cannock Chase Colliery Railway. Rail traffic ceased from 27/9/1963. Locomotives were not used underground.

Gauge: 4ft 8½in

	HANBURY	0-6-0ST	IC	P	567	1894	(a)	(1)
2		0-6-0T	IC	K	5358	1921	(a)	(2)
	McCLEAN	0-4-2ST	IC	BP	28	1856	(b)	(3)
	FOGGO	0-4-2ST	IC	Chasetown		1946	(b)	(4)
	CONDUIT No.4	0-6-0ST	IC	MW	1326	1891	(c)	(5)
	NUTTALL	0-6-0ST	OC	HE	1685	1931	(d)	(6)
		0-4-0ST	OC	AB	2247	1948	(e)	(7)
	THE COLONEL	0-6-0ST	IC	HC	1073	1914	(f)	(8)

(a)   ex Coppice Colliery Co Ltd, 1/1/1947.
(b)   ex Chasetown Loco Shed, after 3/1948).
(c)   ex Conduit Colliery, 10/1949.
(d)   ex Walsall Wood Colliery, between 2/1950 and 6/1950.
(e)   ex Chasetown Loco Shed, between 3/1959 and 6/1959.
(f)   ex Cannock Central Workshops, 11/7/1963.

(1) to Walsall Wood Colliery, 7/1949; ex Cannock Central Workshops, between 4/1954 and 8/1954. To Cannock Central Workshops, between 12/1958 and 4/1959; ex Littleton Colliery between 12/1960 and 4/1961. To Cannock Central Workshops, 2/5/1963.
(2) to Cannock Central Workshops, between 5/1953 and 1/1954, returned between 3/1955 and 5/1955. To Cannock Central Workshops, 13/7/1963.
(3) to Chasetown Loco Shed, by 11/1949.
(4) to Chasetown Loco Shed by 11/1949; returned between 1/1954 and 4/1954; to Brereton Colliery, between 4/1954 and 8/1954.
(5) to Hollybank Loco Shed, 5/1950; returned 10/1950; to Cannock Central Workshops between 10/1950 and 8/1951.
(6) to Walsall Wood Colliery, 8/1955.
(7) to Grove Colliery, between 9/1962 and 3/1963; returned 18/4/1963. To Mid Cannock Colliery, 10/1963.
(8) to Granville Colliery, Salop, 17/10/1963.

EAST CANNOCK COLLIERY, Cannock. SC18
Ex East Cannock Colliery Co.Ltd. (SJ997115)

WM2 from 1/1/1947; CLOSED 5/1957.

Sidings ran east from BR to the colliery at Norton Branch junction (1½ miles north of Cannock Station). An NCB locomotive was used from 1953 only. Locomotives were not used underground.

Gauge: 4ft 8½in

                     4wDM   RH   338413   1953 New (1)

(1) to Cannock Central Workshops, 20/8/1957.

ESSINGTON DISPOSAL POINT, Bloxwich. SF10/12
(1958-70 Site) **(SF10)** SJ986045
(from 1981) **(SF12)** SJ986040

Opened by OE 1958; CLOSED 1970; Reopened by OE on new site 1981.

Sidings on the east side of BR, 2 miles south of Wyrley & Cheslyn Hay Station served the disposal point. These were lifted after the 1970 closure. The site was operated by Stephenson Clarke Ltd. A new disposal point including a rapid loading bunker and sidings laid out for BR merry go round working was opened on the west side of BR in 1981. This is about ¼ mile south of the original and occupies the site of Essington Wood sidings (BR). Rail traffic ceased 12/1992.

Gauge: 4ft 8½in

                     4wDM   FH   3837   1958 New (1)

(1) to Cwm Mawr Disposal Point, Carmarthen, between 10/1969 and 7/1972

GROVE COLLIERY, Little Wyrley.   SE13
Ex W.Harrison Ltd.   (SK019060)

WM2 from 1/1/1947; Merged with **WYRLEY No.3 COLLIERY** 1/1952 and coal winding ceased, Washery & rail system CLOSED 6/63 with WYRLEY No.3.

An NCB branch ran west from BR, 1 mile north of Norton Branch Junction (Pelsall) to the colliery (¾ mile). There was a rail served wharf alongside the Cannock Extension Canal at the colliery. (¾m). Coal was brought here from WYRLEY No.3 Colliery **(SE15)** (which see) by a rope worked overland tubway. Locomotives were not used underground at Grove.

Gauge: 4ft 8½in

No.3		0-6-0ST	IC	P	618	1895	(a)	(1)
	THE COLONEL	0-6-0ST	IC	HC	1073	1914	(a)	(2)
	CONDUIT No.3	0-6-0ST	IC	MW	1180#	1890	(b)	(3)
		0-4-0ST	OC	AB	2247	1948	(c)	(4)
	MARQUIS	0-6-0ST	IC	Lill		1867	(d)	(5)

\#   Incorporated parts from, and plates of, MW 1326 from /1957.

(a)   ex William Harrison Ltd., 1/1/1947.
(b)   ex West Cannock No.5 Colliery, between 5/1960 and 7/1961.
(c)   ex Coppice Colliery, between 9/1962 and 1/1963.
(d)   ex Rawnsley Loco Shed, 15/1/1962.

(1)   to Cannock Central Workshops, between 3/1963 and 4/1963.
(2)   to Cannock Central Workshops, between 5/1960 and 6/1961.
(3)   to Rawnsley Loco Shed, 7/1963.
(4)   to Coppice Colliery, 18/4/1963.
(5)   to Cannock Central Workshops, between 3/1963 and 4/1963.

HAWKINS COLLIERY, Cheslyn Hay.   SE4
(Known as CANNOCK OLD COPPICE COLLIERY until c1950)
Ex T.A.Hawkins & Sons Ltd.   (SJ973079)

WM2 from 1/1/1947; CLOSED 4/1960.

An NCB Branch ran west from the BR Wyrley & Churchbridge goods line passing under the main line en route to the colliery (¾ mile). Traffic was handled for the adjacent works of Rosemary Tileries Ltd (SD3). A short narrow gauge tramway ran north from the colliery to the Staffs & Worcs Canal at Walkmill Bridge (SJ 975083). Locomotives were not used underground.

Gauge: 4ft 8½in

	HAWKINS	0-6-0ST	OC	P	809	1900	(a)	(1)
	TONY	0-6-0ST	OC	HL	3460	1921	(a)	(2)

(a)   ex T.A.Hawkins & Sons Ltd, 1/1/1947.

(1)   to Cannock Central Workshops, between 3/1960 and 4/1961.
(2)   to Chasetown Loco Shed, between 11/1955 and 10/1958.

HILTON MAIN & HOLLYBANK COLLIERIES, Comprising:- SF4/7
HILTON MAIN COLLIERY, Featherstone. (SF7) (SJ943043)
HOLLYBANK COLLIERY, Essington. (SF4) (SJ965033)
Ex Hilton Main & Holly Bank Collieries Ltd.

WM2 from 1/1/1947; **HOLLYBANK COLLIERY**, a service shaft for **HILTON MAIN**, recommenced coal production c1948 but CLOSED 12/1952, therafter becoming a pumping shaft. **HILTON MAIN COLLIERY** became part of STF from 26/3/1967; CLOSED 31/1/1969.

Hollybank Colliery and loco shed were reached by a short NCB line which ran west from exchange sidings (SJ 972033) at the end of a 1¼ mile BR branch from Essington Wood Sidings (1¼ miles north of Bloxwich Station). A longer NCB branch ran south west then north west from a point east of Hollybank Colliery to Hilton Main Colliery (3 miles). Another NCB branch ran east and then south from the exchange sidings to a wharf (SJ973006) on the Wyrley and Essington Canal at Coltham (2 miles). This closed 6/1965. The loco shed at Hollybank remained in use until a new one for the working locos was built at Hilton Main in 1961. The Hollybank shed was demolished in 1963 although locos were stored on the site for some time afterwards. Narrow gauge locomotives were used underground at Hilton Main from 1954 and in the surface stockyard there from 1962.

Gauge: 4ft 8½in

HOLLYBANK No.3		0-6-0ST	IC	HE	1451	1921	(a)	(1)
ROBERT NELSON No.4		0-6-0ST	IC	HE	1800	1936	(a)	(2)
CAROL ANN No.5		0-6-0ST	IC	HE	1821	1936	(a)	(3)
No.1		0-6-0T	IC	HC	352	1891	(a)	(4)
	CONDUIT No.4	0-6-0ST	IC	MW	1326	1896	(b)	(5)
No.2		0-6-0ST	IC	HC	1752	1943	(c)	(6)
No.4		0-6-0DM		WB	3122	1957	New	(7)
No.5		0-6-0DM		WB	3123	1957	(d)	(8)
No.6		0-6-0DE		YE	2748	1959	New	(9)
No.7		0-6-0DE		YE	2749	1959	New	(10)
No.1		0-6-0DM		WB	3117	1956	(e)	(11)
No.2		0-6-0DM		WB	3118	1956	(f)	(12)
No.2		0-6-0ST	IC	HE	3772	1952	(g)	(13)
	NUTTALL	0-6-0ST	OC	HE	1685	1931	(h)	(14)
	HANBURY	0-6-0ST	IC	P	567#	1894	(i)	(15)
No.8		0-6-0ST	IC	HE	3776	1952	(j)	(16)
		4wDM		RH	338413	1953	(k)	(17)

\# Incorporated parts of P 618 from /1963.

(a) ex Hilton Main and Holly Bank Collieries Ltd., 1/1/1947.
(b) ex Coppice Colliery, 5/1950.
(c) ex WD 75091 10/1950.
(d) ex Littleton Colliery, 3/4/1959.
(e) ex West Cannock 1-3 Colliery, 8/1/1959.
(f) ex Littleton Colliery, between 8/1958 and 2/1959.
(g) ex Cannock Central Workshops, between 5/1962 and 7/1962.
(h) ex Cannock Central Workshops, 26/10/1962.
(i) ex Cannock Central Workshops, 3/4/1964.
(j) ex Baggeridge Colliery, between 8/1967 and 2/1968.
(k) ex Baggeridge Colliery, c2/1969.

(1) to Littleton Colliery, 13/1/1959.
(2) to Littleton Colliery, 10/1/1959.
(3) to Littleton Colliery, 9/11/1959.
(4) Scr on site, 11/1958.
(5) to Coppice Colliery, 10/1950.
(6) to Littleton Colliery, 12/1/1959
(7) to Lea Hall Colliery, between 8/1965 and 8/1966.
(8) to West Cannock No 5 Colliery, between 2/1969 and 4/1969.
(9) to Littleton Colliery, 8/ 7/1966.
(10) to Littleton Colliery, between 2/1969 and 4/1969.
(11) to Lea Hall Colliery, 22/4/1960.
(12) to Littleton Colliery, between 11/1959 and 3/1960.
(13) to Littleton Colliery, between 7/1962 and 4/1963.
(14) to Rawnsley Loco Shed, between 10/1962 and 3/1963. Ex Littleton Colliery between 6/1963 and 1/1964. To Lea Hall Colliery, 1/10/1964.
(15) to West Cannock No 5 Colliery, 28/9/1964.
(16) to BR Tyseley Depot (tyre turning) , c24/8/1968.   Thence to Granville Colliery, Shropshire, c9/1968.
(17) to Stafford Wagon Works, Trentham, Stoke on Trent, N Staffs, /1970 (by 8/1970).

Hilton Main Colliery stockyard railway

Gauge: 2ft 6in

| | 4wDM | RH | 476112 | 1962 | New | (1) |
| | 4wDM | RH | 7002/0867/3 | 1967 | (a) | (2) |

(a) ex Cannock Central Stores, /1967

(1) to Lea Hall Colliery, 16/5/1968.
(2) to West Cannock No.5 Colliery, /1968 (by 2/1969).

Underground Haulage

Both endless rope haulage and horses used extensively throughout the workings. From 6/1954 locomotives began to be used underground for coal haulage.

Gauge: 2ft 6in

No.1	0-4-0DMF	RH	370546	1954	New	Scr c/1969
No.2	0-4-0DMF	RH	370552	1954	New	Scr c/1969
No.3	0-4-0DMF	RH	374449	1954	New	Scr c/1969
No.4	0-4-0DMF	RH	388773	1955	New	Scr c/1969
	4wBEF	EEV	3768	1966	New	(1)
	4wBEF	GB	2783	1958	(a)	(2)
	4wBEF	GB	2784	1958	(b)	(3)
	4wBEF	GB	2785	1958	(c)	(4)
	4wBEF	EEV	3841	1967	New	(5)

Carries plate 3148 in error

(a) ex Lea Hall Colliery
(b) ex Lea Hall Colliery, 20/5/1966.
(c) ex Littleton Colliery.

(1) to Lea Hall Colliery, 5/1966.
(2) to Littleton Colliery, by 5/1972.
(3) to Littleton Colliery, 14/4/1969.
(4) to Lea Hall Colliery, c/1965.
(5) to Littleton Colliery, 11/1967

LEA HALL COLLIERY, Rugeley. SA7
(SK058170)

Sunk by WM2 from 1954; Opened 19/7/1960; STF from 26/3/1967; WES from 1/4/1974; CEN from 1/1/1989; CLOSED 21/12/1990.

Sidings ran south east from BR, north of Rugeley Town Station to the colliery (1 mile). Most small coal passed direct by conveyor to the adjacent CEGB Rugeley A power station. Considerable quantities of coal from other collieries was brought in by BR to a wagon tippler for blending and feeding to the power station conveyor. NCB locomotives were employed to handle this traffic plus house coal sales and internal traffic. A small rapid loader was installed in the 1970's for any small coal sold by rail but this saw little use. An extensive surface narrow gauge system was put in for the opening of the colliery, serving stockyards and workshops and connecting with the shaft tops. The locomotives used on the later narrow gauge system shared the standard gauge shed at SK 060168. This system superseded an earlier narrow gauge construction line used during shaft sinking. Locomotives were used extensively underground, mainly for materials and manriding from 1958. The Clayton 'Pony' rubber tyred battery loco, which has since been widely used throughout the NCB for materials haulage on steep gradients, was developed at this colliery between 1964 and 1969.

Gauge: 4ft 8½in

5	BEAUDESERT	0-6-0ST	OC	FW	266	1875	(a)	(1)
	NUTTALL	0-6-0ST	OC	HE	1685	1931	(b)	(2)
No.1	63 000 311	0-6-0DM		WB	3117	1956	(c)	(3)
No.2	63 000 312	0-6-0DM		WB	3118	1957	(d)	(4)
No.4		0-6-0DM		WB	3122	1957	(e)	(5)
No.2	63 000 305	0-6-0DH		EEV	D1120	1966	(f)	(8)
4D		0-6-0DH		HE	7015	1971	New	(6)
No 3D	63 000 316	0-6-0DH		HE	7181	1970	(g)	(8)
	WESTERN JUBILEE							
	63 000 444	6wDE		GECT	5478	1978	(h)	(7)
		0-6-0DH		RR	10240	1965	(i)	(8)
		0-6-0DH		RR	10255	1966	(j)	(8)

(a) ex Rawnsley Loco Shed, between 9/1960 and 7/1961.
(b) ex Hilton Main Colliery, 1/10/1964.
(c) ex Hollybank Loco Shed, 22/4/1960.
(d) ex Littleton Colliery, between 5/1960 and 4/1961.
(e) ex Hilton Main Colliery, between 8/1965 and 8/1966.
(f) ex Littleton Colliery, 4/7/1974. To Walkden Central Workshops, 24/6/1980, returned 7/5/1981.
(g) ex Littleton Colliery, 8/3/1978.

(h) ex Hem Heath Colliery, N Staffs, 29/11/1981.
(i) ex Baddesley Colliery, Warwickshire, between 8/1989 and 10/1989.
(j) ex Baddesley Colliery 7/6/1989.

(1) to Rawnsley Loco Shed, between 7/1961 and 5/1962.
(2) scr on site 8/1966.
(3) to Marple & Gillott Ltd, Sheffield, for scrap, 20/12/1983.
(4) Scr between 10/1978 and 7/1979
(5) to Norton Colliery, 2/9/1971.
(6) to Littleton Colliery, 4/7/1974.
(7) to Coventry Colliery, West Midlands, 24/5/1989.
(8) to Universal Reclamation Ltd, Wellington, Salop, c11/1991.

Construction line for shaft sinking

Gauge: 2ft 6in

| | | 4wDM | RH | 223747 | 1944 | (a) | (1) |

(a) ex Geo Cohen, Sons & Co Ltd, Kingsbury, Warwicks, c/1954

(1) to South Hetton Colliery, Co Durham, 5/1956

Note: There was probably a second RH used here, from the same source.

Stockyard railway

Gauge: 2ft 6in

	63.000.313	4wDM	RH	441947	1959	New	(9)
	63.000.306	4wDM	RH	466587	1961	(a)	(1)
	63.000.310	4wDM	RH	476112	1962	(b)	(6)
LP 61/1	JUMBO	4wDM	RH	506491	1964	(c)	(2)
	63.000.309	4wDM	RH	7002/0767/6	1967	New	(3)
15/24	5074	4wBEF	CE	5074	1966	(d)	(4)
15/18		4wBEF	CE	4960	1964	(e)	s/s 7/1978
		4wDH	HE	8826	1978	New	(8)
	63.000.449	4wDH	HE	8973	1979	New	(5)
	IVOR 2	4wDH	HE	8825	1978	(f)	(7)
	63.000.364	4wDM	RH	441948	1959	(g)	(10)
15/19		4wBEF	CE	5097	1966	(h)	(4)
15/26		4wBEF	GB	6090	1963	(i)	(4)

(a) ex Cannock Wood Colliery, by 4/1962
(b) ex Hilton Main Colliery, 16/5/1969.
(c) ex Highley Colliery, Shropshire, c3/1969.
(d) previously used underground. To surface use, 10/1973.
(e) previously used underground. To surface use between /1969 and /1974.
(f) ex West Cannock No.5 Colliery, 21/6/1983.
(g) ex Wolstanton Colliery, N Staffs, 4/1986.
(h) previously used underground. To surface use, 4/ 6/1988.
(i) ex underground use, /1989; to Ashington National Workshops, 17/5/1989, returned, 31/10/1989

(1) to Cannock Central Workshops, 4/1963; returned c/1964; to Littleton Colliery between /1967 and 3/1969.
(2) to Hem Heath Colliery, Stoke on Trent, 6/1972.
(3) to West Cannock No.5 Colliery, c6/1983.
(4) to Chatterley Whitfield Mining Museum, Stoke on Trent, between 14/6/1991 and 27/8/1991.
(5) to HE, between 11/1979 and 2/1980; returned by 9/1980; to Leicestershire Museums, Snibston Discovery Park, 11/1991
(6) to Walkden Central Workshops, 17/7/1974, returned 28/7/1974. To Walkden CW, 13/5/1978, returned 16/5/1978. To Philadelphia CW, Durham, 5/5/1983, returned 28/11/1984. To Staffs NGRS, Amerton Working Farm, Stowe-by-Chartley, N Staffs, between 3/2/1991 and 27/8/1991.
(7) to Parkside Colliery, Merseyside, between 27/8/1991 and 12/1991
(8) s/s between 3/2/1991 and 27/2/1991
(9) to Cannock Central Workshops, 27/12/1961, returned 22/2/1962. To Wyrley No.3 Colliery, 23/2/1962, returned 16/8/1962. To ? (repairs), 23/12/1972; returned /1973.
Written off 2/1982, and Scr between 11/4/1986 and 19/1/1987
(10) To Staffs NGRS, Amerton Working Farm, Stowe-by-Chartley, N Staffs, between 3/2/1991 and 27/8/1991.

Underground Locomotives

Gauge: 2ft 6in

No 5		4wBEF	GB	2783	1958	New	(1)
No 4		4wBEF	GB	2784	1958	New	(2)
15/3	No 3	4wBEF	GB	2785	1958	New	(3)
15/2		4wBEF	GB	2786	1959	New (l)	(18)
15/8		4wBEF	EE	2846	1960		
			RSHN	8133	1960	(a)	(18)
15/7		4wBEF	EE	2861	1960		
			RSHD	8204	1960	New (m)	(18)
15/4		4wBEF	GB	2901	1959	(b)	(4)
15/1		4wBEF	GB	2987	1960	(c)	(5)
15/9		4wBEF	EE	3146	1961		
			RSHD	8288	1961	New (n)	(18)
15/10		4wBEF	EE	3147	1961		
			RSHD	8289	1961	New	(14)
15/11		4wBEF	EE	3223	1962		
			RSHD	8344	1962	New (o)	(16)
15/12		4wBEF	EE	3224	1962		
			RSHD	8345	1962	New (p)	(15)
15/13		4wBEF	EE	3400	1963		
			RSHD	8420	1963	New (q)	(17)
15/26		4wBEF	GB	6090	1963	(d)	(6)
15/14		4wBEF	GB	6091	1963	New	(18)
		4wBEF	CE	4727	1964	New	s/s by 7/1978
15/15		4wBEF	CE	4805	1964	New	Scr c/1984
15/18		4wBEF	CE	4960	1964	New	(7)
15/16		4wBEF	EEV	3493	1964	New	(17)
15/17		4wBEF	EEV	3494	1964	New (r)	(18)
15/24		4wBEF	CE	5074	1966	(e)	(8)

15/19		4wBEF	CE	5097	1966	New	(9)
15/20		4wBEF	EEV	3768	1966	(f) (s)	(14)
		4wBEF	EEV	3840	1967	New	(12)
15/21		4wBEF	EEV	3995	1971	New (t)	(17)
15/22		4wBEF	CE	5896A	1972	New	(10)
15/23		4wBEF	CE	5896B	1972	New	s/s c/1986
15/24		4wBEF	CE	5962A	1973	New	(18)
15/25		4wBEF	CE	5962B	1973	New	s/s c/1988
15/26		4wBEF	CE	B0909A	1976	New	(10)
15/27		4wBEF	CE	B0909B	1976	New	(13)
		4wBEF	EE	1810	1952		
			Bg	3354	1952	(g)	Scr c/1984
15/29		4wBEF	GECT	5571	1978	New	(17)
15/30		4wBEF	GECT	5572	1978	New	(18)
15/28		4wBEF	CE	B1828B	1979	(h)	(11)
15/31		4wBEF	CE	B1886B	1980	New	(18)
15/5		4wBEF	EE	3151	1961		
			RSHD	8291	1961	(i)	(18)
8/5	15/1		EE	3402	1963		
			RSHD	8421	1963	(j)	(15)
1		4wBEF	Bg	3554	1961	(k)	(18)
15/15		4wBEF	GECT	5419	1977	(u)	(18)

(a) ex Cannock Wood Colliery 8/1960. To Walkden Central Workshops, Gtr Manchester, 15/7/1980, returned 28/10/1980. To Philadelphia Central Workshops, Co Durham, 23/11/1982; returned 7/7/1983.
(b) ex Littleton Colliery, c/1960.
(c) ex Littleton Colliery, 28/9/1971.
(d) ex Cannock Wood Colliery, 1/1974.
(e) ex Valley Training Centre, Hednesford, 10/7/1969.
(f) ex Hilton Main Colliery, 5/1966.
(g) ex West Cannock No.5 Colliery, 8/1977.
(h) ex Littleton Colliery, 6/4/1979.
(i) ex Mining Research & Development Establishment, Swadlincote, Derbys, c/1983.
(j) ex Philadelphia Central Workshops, Co Durham, 1/1985, prev West Cannock No.5 Colliery.
(k) ex Wolstanton Colliery, N Staffs, c/1986.
(l) to Walkden Central Workshops, Gtr Manchester, 8/1976. Returned 11/1976.
(m) to Philadelphia Central Workshops, Co Durham, 12/1982. Returned 6/1983.
(n) to Walkden Central Workshops, Gtr Manchester, c/1978. Returned 8/1978.
(o) to Walkden Central Workshops, Gtr Manchester, 9/11/1978. Returned 6/2/1980
(p) to Philadelphia Central Workshops, Co Durham, 18/2/1985, returned 13/2/1986.
(q) to Walkden Central Workshops, Gtr Manchester, 11/1981. Returned 12/1982.
(r) to Walkden Central Workshops, Gtr Manchester, 19/3/1980. Returned 16/3/1981. To Philadelphia Central Workshops, Co Durham, c/1982, returned c/1983.
(s) to Walkden Central Workshops, Gtr Manchester, c/1972. Returned 1/1974. To Ashington National Workshops, Northumberland, 3/1986, returned 11/1988.
(t) to Walkden Central Workshops, Gtr Manchester, 10/3/1976, returned 22/4/1976. To Walkden again 24/8/1981, returned 15/6/1982.
(u) ex Philadelphia Central Workshops, Co Durham, 9/1983 (prev West Cannock No.5 Colliery)

(1) to Hilton Main Colliery.
(2) to Hilton Main Colliery, 20/5/1966
(3) to Littleton Colliery, by 1/1965; ex Hilton Main Colliery, c/1965. To Walkden Central Workshops, Gtr Manchester, c/1970; returned c10/1970. Scr between 28/1/1986 and 19/1/1989.
(4) to West Cannock No.5 Colliery, 5/1961.
(5) to Walkden Central Workshops, Gtr Manchester, by 8/1972, returned 12/1972. Scr between 28/1/1986 and 19/1/1989.
(6) transferred to stockyard railway, /1989.
(7) transferred to stockyard railway, between /1969 and /1974.
(8) transferred to stockyard railway, 10/1973.
(9) transferred to stockyard railway, 14/6/1988.
(10) to Florence Colliery, N Staffs, 12/1986.
(11) to Littleton Colliery, c10/1989.
(12) to West Cannock No.5 Colliery, 4/1967
(13) to Bolsover Colliery, Derbyshire, /1991
(14) to Harworth Colliery, Notts, 2/8/1991
(15) to Cotgrave Colliery, Notts, 5/8/1991
(16) to Chatterley Whitfield Mining Museum, Tunstall, N Staffs, c11/1991
(17) to Shropshire Locomotive Collection, c3/1992
(18) abandoned underground after closure of colliery, 12/1990.

LITTLETON COLLIERY, Huntington. SB2
Ex Littleton Collieries Ltd. (SJ973127)

WM2 from 1/1/1947; STF from 26/3/1967; WES from 1/4/1974; CEN from 1/1/1989; MWG from 1/10/1991.

An NCB branch runs south east from BR, ½ mile south of Penkridge station to Otherton (SJ 933119) (1¼ miles). Sidings here formerly served a canal wharf and are now used for empty wagon stabling. The branch continues east climbing steeply to the colliery (4 miles). A rapid loading bunker was installed c1977 but because the steep branch could not readily be brought up up to BR standards, haulage has remained in the hands of NCB locomotives. Trackwork was then rationalised and other rail traffic thereafter consisted of small tonnages of house coal only. A private warehouse between Otherton and Penkridge is rail connected to the NCB line and its traffic is worked as required. A narrow gauge surface rail system in the stockyard has been locomotive worked since 1959 and is unusual in that locomotives are repaired on a standard gauge wagon by which they are conveyed from the stockyard as necessary. Locomotives have been used underground for coal haulage, manriding and materials haulage since 1950, though conveyors were later introduced for coal transport.

Gauge: 4ft 8½in

	LITTLETON No.1	0-6-0ST	IC	MW	1515	1901	(a)	(1)
	LITTLETON No.2	0-6-0ST	IC	MW	1596	1903	(a)	(2)
	LITTLETON No.4	0-6-0ST	IC	MW	1759	1910	(a)	Scr 9/1961.
	LITTLETON No.5	0-6-0ST	IC	MW	2018	1922	(a)	(3)
	CONDUIT No.3	0-6-0ST	IC	MW	1180	1890	(a)	(4)
	ADJUTANT	0-6-0ST	OC	MW	1913	1917	(b)	(5)
No.2		0-6-0DM		WB	3118	1957	(c)	(6)
	LITTLETON No.6	0-6-0ST	IC	RSH	7292	1945	(d)	(7)
No.7		0-6-0ST	IC	HC	1752	1943	(e)	(8)
	HOLLYBANK No.3	0-6-0ST	IC	HE	1451	1924	(f)	(9)

	ROBERT NELSON No.4	0-6-0ST	IC	HE	1800	1936	(g)	(3)
3	HANBURY	0-6-0ST	IC	P	567	1894	(h)	(10)
	CAROL ANN No.1							
	(CAROL ANN No.5)	0-6-0ST	IC	HE	1821	1936	(i)	(11)
No.2		0-6-0ST	IC	HE	3772	1952	(j)	(12)
No.8		0-6-0ST	IC	RSH	7106	1943	(k)	(13)
	NUTTALL	0-6-0ST	OC	HE	1685	1931	(l)	(14)
No.6	63000320	0-6-0DE		YE	2748	1959	(m)	(15)
No.7	63000319	0-6-0DE		YE	2749	1959	(n)	(16)
No.5		0-6-0DM		WB	3123	1957	New	(17)
3D		0-6-0DH		HE	7181	1970	New	(18)
8D		0-6-0DH		HE	7018	1971	New	(19)
No.2		0-6-0DH		EEV	D1120	1966	(o)	(20)
4D		0-6-0DH		HE	7015	1971	(p)	(21)
63 000 441								
	WESTERN ENTERPRISE	6wDE		GECT	5421	1977	New	
63 000 442								
	WESTERN PIONEER	6wDE		GECT	5422	1977	New	(22)
63 000 443								
	WESTERN PROGRESS	6wDE		GECT	5468	1977	(q)	
63 000 445								
	WESTERN QUEEN	6wDE		GECT	5479	1979	(r)	
63 000 446								
	WESTERN KING	6wDE		GECT	5480	1979	(r)	

(a) ex Littleton Collieries Ltd., 1/1/1947.
(b) on hire at vesting day (1/1/1947) from Cannock & Rugeley Colliery Co Ltd.
(c) ex West Cannock No 1-3 Collieries, between 7/1957 and 10/1957.
(d) ex WD 71483, 5/1947.
(e) ex Hollybank Loco Shed, 12/1/1959.
(f) ex Hollybank Loco Shed, 13/1/1959
(g) ex Hollybank Loco Shed, 10/1/1959
(h) ex Cannock Central Workshops, between 3/1960 and 12/1960.
(i) ex Hollybank Loco Shed, 9/11/1959.
(j) ex Hollybank Loco Shed, between 7/1962 and 4/1963.
(k) ex Cannock Central Workshops, 19/4/1963.
(l) ex Rawnsley Loco Shed, 6/1963.
(m) ex Hilton Main Colliery, 8/7/1966.
(n) ex Hilton Main Colliery, between 2/1969 and 4/1969.
(o) ex Cannock Wood Colliery, 11/10/1973.
(p) ex Lea Hall Colliery, 4/7/1974.
(q) ex Hem Heath Colliery, 15/12/1977; to GECT 3/3/1978; returned 8/6/1978; to British Steel plc, Shap Works, Cumbria, 9/1990 (hire) en route to RFS Doncaster, S Yorks for tyre turning. Returned from RFS 1/1990.
(r) ex Bickershaw Colliery, 3/1987.

(1) to Cannock Central Workshops, between 2/1954 and 9/1955; returned between 11/1955 and 8/1957; Scr on site 9/1961.
(2) Scr on site c14/11/1960.
(3) to Foxfield RPS, Dilhorne, N Staffordshire, 10/1972.
(4) to West Cannock No.5 Colliery, 10/1949.
(5) to Rawnsley Shed, /1947.
(6) to Hollybank Loco Shed, between 8/1958 and 2/1959; returned between 11/1959 and 3/1960; to Lea Hall Colliery, between 5/1960 and 4/1961..

(7) to WB for repairs between 2/1954 and 5/1954, returned c/1954. To Cannock Central Workshops, between 12/1957 and 4/1958, returned between 12/1958 and 3/1959. Scr on site, 10/1970.
(8) to Bold Colliery, Lancashire, 27/2/1978.
(9) to Cannock Central Workshops, between 3/1959 and 6/1959; returned between 5/1960 and 6/1961; to Granville Colliery, Shropshire, 9/1966.
(10) to Coppice Colliery, between 12/1960 and 4/1961.
(11) scrapped by Thos.W.Ward Ltd on site, 8/1966.
(12) scrapped on site, 10/1969.
(13) scrapped on site, /1968.
(14) to Hilton Main Colliery, between 6/1963 and 1/1964.
(15) to Peak Rail, Buxton, Derbys, 14/7/1988.
(16) to Walkden Central Workshops, Gtr Manchester, 5/1978. To Bold Colliery, Lancs 4/1979.
(17) to Bankfoot Loco Shed, Co Durham (demonstration), 9/1958; returned 10/1958; to Hollybank Loco Shed, 3/4/1959.
(18) to Lea Hall Colliery, 8/3/1978.
(19) to Hem Heath Colliery, between 9/1977 and 4/1978.
(20) to Lea Hall Colliery, 4/7/1974.
(21) to Hem Heath Colliery, between 7/1977 and 1/1978.
(22) to GECT, 10/1/1978, returned 14/2/1978. To Walkden Central Workshops, 9/5/1978, returned 15/6/1978. To Coventry Colliery 22/5/1989.

Stockyard railway

Gauge: 2ft 6in

	63000315		4wDM	RH	441946	1959		
				Carries plate	441944 in error	New	(1)	
	63000306		4wDM	RH	466587	1961	(a)	(2)
			4wDM	RH	452293	1960	(b)	(3)
	63000348	LP61	4wDH	RH	476107	1964	(c)	(4)
15			4wDM	RH	506491	1964	(d)	
No.7	63000447		4wDH	HE	8971	1979	New	
No.8	COMET		4wDH	HE	9041	1982	New	

(a) ex Lea Hall Colliery, between /1967 and 3/1969.
(b) ex Cannock Wood Colliery, between 12/1970 and 10/1971.
(c) ex Hem Heath Colliery, N Staffs, (possibly via Trentham Machinery Stores), 16/6/1975.
(d) ex Trentham Machinery Stores, N Staffs, 19/8/1975. to Walkden Central Workshops, Gtr Manchester, between 23/2/1979 and 2/3/1979. Returned 16/1/1980.

(1) to Walkden Central Workshops, Gtr Manchester, between 2/1974 and 6/1975. To West Cannock No.5 Colliery, c1/1977.
(2) to Cannock Wood Colliery, between 10/1969 and 10/1971. Returned from Trentham Machinery Stores, 3/5/1974; to Walkden Central Workshops, 12/2/1975.
(3) to Walkden Central Workshops, Gtr Manchester, or Trentham Machinery Store, and s/s c/1975.
(4) to Walkden Central Workshops, 11/9/1981. Written off and remains returned to Littleton Colliery, 5/3/1982, where Scr on site, 3/1982.

Underground locomotives.

When this colliery passed to the NCB endless ropes and pit ponies were used for haulage underground. Locomotives were introduced in 1950 in both No.2 and No.3 pits for coal and material haulage from points up to 1800 yds from the pit bottom..

Gauge: 2ft 6in

No.1			0-4-0DMF	HE	4080	1950	New	(2)
No.2			0-4-0DMF	HE	4081	1950	New	(1)
No.3			0-4-0DMF	HE	4082	1950	New	(3)
No.4			0-4-0DMF	HE	4083	1950	New	(2)
No.5			0-4-0DMF	HE	4084	1950	New	(3)
No.6			0-4-0DMF	HE	4085	1951	New	(4)
No.7			0-4-0DMF	HE	4086	1951	New	(5)
11/3			4wBEF	EE	1810	1952		
				Bg	3354	1952	New	(6)
No.1	V1/1	11/2	4wBEF	EE	2296	1955		
				Bg	3454	1955	New	
No.2	V1/2		4wBEF	EE	2298	1955		
				Bg	3456	1955	(a)	
No.1	V1/1		4wBEF	GB	2745	1956	New	
No.5	V1/5	11/9	4wBEF	GB	2783	1958	(b)	
No.4			4wBEF	GB	2784	1958	(c)	
No.6			4wBEF	GB	2785	1958	(d)	(7)
No.3	V1/3		4wBEF	EE	2660	1959		
				RSHN	7945	1959	New	
No.4	V1/4		4wBEF	EE	2741	1959		
				RSHN	8129	1959	New	(8)
			4wBEF	GB	2901	1959	New	(9)
No.14	V1/14		4wBEF	EE	2844	1960		
				RSHN	8131	1960	(e)	(10)
No.15	V1/15		4wBEF	EE	2860	1960		
				RSHD	8203	1960	(f)	
No.5	V1/5		4wBEF	EE	2862	1960		
				RSHD	8205	1960	New (k)	
No.6	V1/6		4wBEF	EE	2927	1960		
				RSHD	8206	1960	New	
No.2	V1/2		4wBEF	GB	2987	1960	New	(11)
No.3	V1/3		4wBEF	GB	2988	1960	New (l)	
No.7			4wBEF	EE	3151	1961		
				RSHD	8291	1961	New	(12)
No.8	V1/8		4wBEF	EE	3152	1961		
				RSHD	8292	1961	New (n)	
No.9	V1/9		4wBEF	EEV	3652	1965	New	
No.10	V1/10		4wBEF	EEV	3769	1969	New (o)	
No.11			4wBEF	EEV	3841	1970	(g)	
No.12	V1/12		4wBEF	Bg	3582	1962	(h)	
No.1			4wBEF	CE	B1828A	1979	New	
			4wBEF	CE	B1828B	1979	New (p)	
No.2			4wBEF	CE	B1886A	1980	New	
No.3			4wBEF	CE	B1894	1980	New	
No.4			4wBEF	CE	B2927	1981	New	
No.16	V1/16		4wBEF	EE	3157	1961		
				RSHD	8297	1961	(i)	
			4wBEF	CE	B3428	1988	(j)	

(a) ex Wyrley No.3 Colliery, 9/5/1963.
(b) ex Hilton Main Colliery, by 5/1972.
(c) ex Hilton Main Colliery, 14/4/1969.
(d) ex Lea Hall Colliery, by 1/1965.
(e) ex Cannock Wood Colliery, 11/1972.
(f) ex Cannock Wood Colliery, 28/6/1973; to Walkden Central Workshops, 26/3/1974; retd 17/3/1975.
(g) ex Hilton Main Colliery, 11/1967. To Walkden Central Workshops, Gtr Manchester, c/1978, returned 10/1978.
(h) ex Wolstanton Colliery, 11 or 14/2/1968. To Philadelphia Central Workshops, Co Durham, 11/1983, returned 19/9/1985.
(i) ex West Cannock No.5 Colliery, 26/8/1982.
(j) ex Sherwood Colliery, Nottinghamshire /1990 (2ft 3in gauge). To CE, /1990 (2ft 6in gauge), returned /1990.
(k) to Walkden Central Workshops, Gtr Manchester, 18/12/1978, returned 7/1980.
(l) to Walkden Central Workshops, Gtr Manchester, 25/10/1972, returned 1/5/1975.
(n) to Walkden Central Workshops, Gtr Manchester, 5/1979, returned 10/1980.
(o) to Walkden Central Workshops, Gtr Manchester, 10/11/1977, returned 10/1978.
(p) to Lea Hall Colliery, 6/4/1979, returned c10/1989.

(1) to Mid Cannock Colliery, 8/1953.
(2) to Mid Cannock Colliery, after /1956.
(3) to West Cannock No.5 Colliery, after /1956.
(4) abandoned underground, c/1972.
(5) to Trentham Machinery Store, N Staffs, by /1978.
(6) to Wyrley No.3 Colliery, before /1963.
(7) to Hilton Main Colliery.
(8) to Walkden Central Workshops, Gtr Manchester, 1/1983.
(9) to Lea Hall Colliery, c/1960.
(10) to Florence Colliery, N Staffs, 7/1984.
(11) to Lea Hall Colliery, 28/9/1971.
(12) to Walkden Central Workshops, Gtr Manchester, 12/1980, returned 10/1981. To Mining Research & Development Establishment, Swadlincote Test Site, Derbys, /1982.

Becorit Road Rail System

Gauge: 200mm

- 1adDMF BGB 25/2/216 1971 (a) (1)

(a) delivered new to Littleton, but not used or sent underground.

(1) to Valley Training Centre, /1971

MID CANNOCK COLLIERY & DISPOSAL POINT, Cannock. SD2
Ex W.Harrison Ltd. (SJ987091)

WM2 from 1/1/1947; STF from 26/3/1967; CLOSED 12/1967 and site cleared.
Reopened as a Disposal Point by OE 1979; On care and maintenance basis from c1988.

Sidings on the east side of the BR line, south of Cannock Station served the colliery. Until NCB locomotives were introduced in 1958, these were worked by BR, supplemented by a number of electric haulages or capstans. A rope hauled narrow gauge tubway ran east from the colliery to a wharf on the Cannock Extension Canal. This seems to have been removed by 1960. In 1958 a locomotive was introduced in the suface stockyard. Locomotives were introduced underground in 8/1953 in preparation for the merger of Cannock & Leacroft Colliery into Mid-Cannock. They were used for coal and materials haulage and manriding. , The new disposal point on the site is served by sidings equipped for rapid loading with haulage by BR locomotives throughout .

Gauge: 4ft 8½in

No.3   0-6-0ST   OC   P   879   1901   (a)   (1)
                     4wDM   RH   338413 1953 (b)   (2)
       STAFFORD   0-6-0T   IC   HC   319   1889   (c)   (3)
                     0-4-0ST   OC   AB   2247 1948   (d)   (4)
                     4wDM   RH   321730 1952 (e)   (5)

(a)   ex West Cannock 1-3 Colliery, between 11/1957 and 12/1958.
(b)   ex Cannock Central Workshops, 5/1958
(c)   ex West Cannock 1-3 Colliery, between 6/1959 and 2/1960.
(d)   ex Coppice Colliery, 10/1963.
(e)   ex Baggeridge Colliery, between 10/1963 and 4/1964.

(1)   Scr on site by W.H.Arnott Young & Co Ltd, c8/1961.
(2)   to Baggeridge Colliery, between 3/1967 and 3/1968.
(3)   to West Cannock 1-3 Colliery, 4/1960.
(4)   to Walsall Wood Colliery, 5/1964.
(5)   to Parkhouse Colliery, N Staffs, 12/4/1968.

Stockyard Railway

An extensive railway was laid around the stockyard which lay to the north of the shafts.

Gauge 2ft6in

              4wDM   RH   441948   1959
              Carried plate 436862 in error   New   (1)

(1)   to Wolstanton Colliery, N Staffs, between 12/1967 and 3/1968.

Note: The unidentified RH loco from Wyrley No.3 Colliery may have been here for a period after 8/1963.

Underground locomotives

Pit ponies used underground for general haulage duties, but gradually replaced by locomotives from c/1953. In 1963 Wyrley No.3 pit workings were merged with those of Mid Cannock and Mid Cannock locomotives thereafter hauled coal over the combined underground system to be drawn at the Mid Cannock shaft.

Gauge: 2ft 6in

	0-4-0DMF	HE	4080	1950	(b)	(1)
	0-4-0DMF	HE	4081	1950	(a)	(1)
	0-4-0DMF	HE	4083	1950	(b)	(1)
	0-4-0DMF	HE	4087	1952	(c)	(1)
	0-4-0DMF	HE	4497	1956	(c)	(1)
No.1	0-4-0DMF	RH	339267	1953	New	(2)
No.2	0-4-0DMF	RH	339268	1953	New	(1)
No.3	0-4-0DMF	RH	339275	1953	New	(1)
No.4	0-4-0DMF	RH	374452	1954	New	(1)
No.5	0-4-0DMF	RH	392155	1956	New	(1)

(a) ex Littleton Colliery, 8/1953
(b) ex Littleton Colliery, after /1956.
(c) ex West Cannock No.5 Colliery.

(1) to Atherstone Iron & Steel Co Ltd, Warwicks, for scrap, c/1968.
(2) to Wolstanton Colliery, N Staffs, 2/1968.

NOOK & WYRLEY COLLIERY, Cheslyn Hay.   SD1
Ex Nook & Wyrley Collieries Ltd.   (SJ986068)

WM2 from 1/1/1947; CLOSED 6/1949.

Served by sidings on the west side of BR, ½ mile south of Wyrley & Cheslyn Hay station. Locomotives were not used underground.

OLD COPPICE COLLIERY - see HAWKINS

RAWNSLEY LOCO SHED - see CANNOCK WOOD COLLIERY.

VALLEY COLLIERY & TRAINING CENTRE.-
see CANNOCK & RUGELEY COLLIERIES.

WEST CANNOCK No.1, 2 & 3 COLLIERIES, Comprising:- SC12
WEST CANNOCK No.1 & 2 COLLIERIES, Hednesford. (SC12) (SJ993129)
WEST CANNOCK No.3 COLLIERY, Hednesford. (SC14) (SJ995124)
Ex West Cannock Colliery Co.Ltd.

WM2 from 1/1/1947; **No.3 Colliery** CLOSED 12/1949; **No.2 Colliery** merged into No.5 in 1/1955; No.1 merged into **LITTLETON** 9/1958; Surfaces CLOSED by 1961.

An NCB branch ran west to No.3 Colliery (½ mile), then north west to Nos.1 & 2 Colliery (¾ mile) from BR, south of Hednesford station. The locomotive shed was located in the exchange sidings at SJ998123. A surface rope hauled narrow gauge tubway ran south east from the three collieries, to a wharf on the Cannock Extension Canal (SJ 996111) (1 mile), south of East Cannock Colliery and near to that of the Cannock & Rugeley Collieries. Locomotives were not used underground.

Gauge: 4ft 8½in

	STAFFORD	0-6-0T	IC	HC	319	1889	(a)	(1)
	TOPHAM	0-6-0ST	OC	WB	2193	1922	(a)	(2)
	CONDUIT No.3	0-6-0ST	IC	MW	1180	1890	(b)	(3)
No.3		0-6-0ST	OC	P	879	1901	(c)	(4)
	NUTTALL	0-6-0ST	OC	HE	1685	1931	(d)	(5)
No.1		0-6-0DM		WB	3117	1957	New	(6)
No.2		0-6-0DM		WB	3118	1957	New	(7)

(a) ex West Cannock Colliery Co Ltd, 1/1/1947.
(b) ex West Cannock No.5 Colliery, Brindley Heath, between 4/1954 and 11/1954.
(c) ex West Cannock No.5 Colliery, between 12/1953 and 11/1954.
(d) ex Cannock Central Workshops, 30/5/1957.

(1) to Cannock Central Workshops, between 3/1957 and 7/1957, returned between 12/1958 and 6/1959. To Mid Cannock Colliery, between 6/1959 and 2/1960, returned 4/1960. To West Cannock No.5 Colliery, between 9/1960 and 11/1961
(2) to West Cannock No.5 Colliery, between 11/1954 and 5/1955, returned between 8/1956 and 11/1956. To West Cannock No.5 Colliery, again, 5/1958.
(3) to Cannock Central Workshops, between 10/1955 and 8/1956.
(4) to Mid Cannock Colliery, between 11/1957 and 12/1958.
(5) to Chasetown Loco Shed, 2/7/1957.
(6) to Hollybank Loco Shed, 8/1/1959.
(7) to Littleton Colliery, between 7/1957 and 10/1957

WEST CANNOCK No.5 COLLIERY, Hednesford. SC16
Ex West Cannock Colliery Co.Ltd. (SK007142)

WM2 from 1/1/1947; STF from 26/3/1967; WES from 1/4/1974; CLOSED 12/1982.

Sidings ran north from BR, 1 mile north of Hednesford Station, to the colliery (¾ mile). The surface was extensively modernised in 1957 including a new coal preparation plant and later a new locoshed at SK007144. Rail traffic ceased from 12/1977, after which coal went by road to Littleton Colliery for washing. A locomotive was used in the surface stockyard from 1959. Locomotives were used underground from 1950 with their use extended c1960 to reach reserves acquired from No.1 and Wimblebury Collieries.

Gauge: 4ft 8½in

	BLACKCOCK	0-4-2ST	IC	BP	1140	1871		
					Reb	1930	(a)	(1)
No.3		0-6-0ST	OC	P	879	1901	(a)	(2)
	CONDUIT No.3	0-6-0ST	IC	MW	1180#	1890	(b)	(3)
	TOPHAM	0-6-0ST	OC	WB	2193	1922	(c)	(4)
	ANGLESEY	0-6-0ST	IC	Lill		1868	(d)	(5)
	NUTTALL	0-6-0ST	OC	HE	1685	1931	(e)	(6)
No.4		0-6-0ST	IC	HE	3806	1953	(f)	(7)
	STAFFORD	0-6-0T	IC	HC	319	1889	(g)	(8)
	HANBURY	0-6-0ST	IC	P	567*	1894	(h)	(9)
No.5	63 000 325	0-6-0DM		WB	3123	1957	(i)	(10)
No.8	63 000 326	0-6-0ST	IC	HE	3776	1952	(j)	(11)
6D	63 000 322	0-6-0DH		HE	7017	1971	New	(12)

\#   Incorporated parts from, and plates of, MW 1326 from 1957
*   Incorporated parts from P 618 from /1963

(a) ex West Cannock Colliery Co Ltd., 1/1/1947.
(b) ex Littleton Colliery, 10/1949.
(c) ex West Cannock No 1-3 Colliery, between 11/1954 and 5/1955.
(d) ex Rawnsley Shed, between 8/1956 and 4/1957.
(e) ex Chasetown Loco Shed, 8/1957.
(f) ex Cannock Chase Loco Shed, between 9/1961 and 1/1962.
(g) ex West Cannock No 1-3 Colliery, between 9/1960 and 11/1961.
(h) ex Hilton Main Colliery, 25/9/1964.
(i) ex Hilton Main Colliery, between 2/1969 and 4/1969.
(j) ex Cannock Wood Colliery, 1/1971.

(1) scrapped by colliery fitters 12/1949.
(2) to West Cannock No 1-3 Colliery, between 12/1953 and 11/1954.
(3) to West Cannock No 1-3 Colliery, between 4/1954 and 11/1954. Returned from Cannock Central Workshops, between 7/1957 and 11/1957. To Grove Colliery, between 5/1960 and 7/1961.
(4) to West Cannock No 1-3 Colliery, between 8/1956 and 11/1956, returned 5/1958. To Cannock Central Workshops, between 5/1960 and 6/1961, returned 7/9/1962. To Cannock Wood Colliery 2/1970, returned 9/1970. To Foxfield RPS, Dilhorne, Staffs, 10/1972.
(5) to Rawnsley Shed, between 4/1957 and 7/1957.
(6) to Cannock Chase Loco Shed, between 11/1957 and 3/1958.
(7) to Dean Forest RPS, Parkend, Gloucestershire, c3/1973.

(8) to Rawnsley Shed, 5/1963.
(9) Scr on site, between 11/1969 and 1/1970.
(10) to Wolstanton Colliery, N Staffs, 7/1977.
(11) to Bickershaw Colliery, Lancashire, 9/2/1977.
(12) to Granville Colliery, Shropshire, between 2/1977 and 6/1978.

Stockyard Railway.

New stockyard brought into use c/1959. Rail traffic ceased 1983.

Gauge: 2ft 6in

	63000323	4wDM	RH 441945	1959	New	(1)
2	63000324	4wDM	RH 7002/0867/3	1967	(a)	(2)
	63000315	4wDM	RH 441946	1959	(b)	(3)
No.3	63000369	4wDM	RH 497760	1963	(c)	(4)
	IVOR 2	4wDH	HE 8825	1978	(d)	(5)
No.2	63000309	4wDM	RH 7002/0767/6	1967	(e)	(6)

(a) ex Hilton Main Colliery, by 2/1969.
(b) ex Trentham Machinery Stores, N Staffs, c1/1977.
(c) ex Chatterley Whitfield Colliery, N Staffs, 20/2/1978.
(d) ex Trentham Machinery Stores, N Staffs, 10/1979.
(e) ex Lea Hall Colliery, c6/1983
(1) to Walkden Central Workshops, Gtr Manchester, by 8/1976 and later Trentham Machinery Stores, N Staffs, 6/1977
(2) to Walkden Central Workshops, Gtr Manchester, by 6/1977 and later Wolstanton Colliery, N Staffs, c1/1978.
(3) to Walkden Central Workshops, Gtr Manchester, 1/11/1979, returned 9/5/1980. To Parkside Colliery, Lancashire, between c/1983 and 7/1985.
(4) to Trentham Machinery Stores, N Staffs, 8/1979
(5) to Lea Hall Colliery, 21/6/1983.
(6) to Chatterley Whitfield Mining Museum, Tunstall, Staffs, between 6/1983 and 9/1984.

Underground locomotives

Gauge: 2ft 6in

	0-4-0DMF	HE	4087	1952	New	(1)
	0-4-0DMF	HE	4088	1952	New	Scr by /1972.
	0-4-0DMF	HE	4082	1950	(a)	(2)
	0-4-0DMF	HE	4084	1950	(a)	Scr by /1972.
	0-4-0DMF	HE	4497	1956	New	(1)
8/2	4wBEF	EE	3156	1961		
		RSHD	8296	1961	New	(3)
8/1	4wBEF	EE	3157	1961		
		RSHD	8297	1961	New	(4)
15/4	4wBEF	GB	2901	1959	(b)	(3)
8/3	4wBEF	EE	3161	1962		
		RSHD	8302	1962	New	(3)
8/4	4wBEF	GB	6076	1962	New	(3)
8/5	4wBEF	EE	3402	1963		
		RSHD	8421	1963	New	(5)

11/2		4wBEF	EE	2297	1955		
			Bg	3455	1955	(c)	(3)
11/3		4wBEF	EE	1810	1952		
			Bg	3354	1952	(c)	(6)
8/6		4wBEF	EEV	3495	1964	New	(7)
5074		4wBEF	CE	5074	1965	New	(8)
		4wBEF	EE	2842	1960		
			Bg	3547	1960	(d)	Scr c/1970
8/7		4wBEF	EEV	3840	1967	(e)	(9)
8/8		4wBEF	EE	2847	1960		
			RSHN	8134	1960	(f)	(10)
8/9		4wBEF	EE	2845	1960		
			RSHN	8132	1960	(g)	(3)
8/10		4wBEF	GECT	5419	1977	New	(11)
8/11		4wBEF	GECT	5420	1977	New	(12)
8/9		4wBEF	GECT	5433	1977	New	(3)

(a) ex Littleton Colliery, after /1956
(b) ex Lea Hall Colliery, 5/1961.
(c) ex Wyrley No.3 Colliery, 10/1963.
(d) ex Highley Colliery, Shropshire, c/1970.
(e) ex Lea Hall Colliery, 4/1967.
(f) ex Cannock Wood Colliery, 7/1973.
(g) ex Cannock Wood Colliery, 11/1973.

(1) to Mid Cannock Colliery.
(2) abandoned underground c/1972.
(3) written off after 12/1982, disposal unknown
(4) to Walkden Workshops, Gtr Manchester, 1/11/1979, returned 24/9/1980. To Littleton Colliery, 26/8/1982.
(5) to Philadelphia Central Workshops, Co Durham, 3/1983. To Lea Hall Colliery, 1/1985
(6) to Lea Hall Colliery 8/1977.
(7) to Hem Heath Colliery 3/1983.
(8) to Valley Training Centre by /1969.
(9) to Walkden Workshops, Gtr Manchester, 3/1978, returned 3/1978. To Hem Heath Colliery, N Staffs, c10/1983.
(10) to Philadelphia Central Workshops, Co Durham, 1/1983, then Agecroft Colliery, Gtr Manchester, c/1983.
(11) to Philadelphia Central Workshops, Co Durham, 1/1983, then Lea Hall Colliery, 9/1983.
(12) to Hem Heath Colliery, N Staffs, 1/1983.

WIMBLEBURY COLLIERY - see CANNOCK & RUGELEY COLLIERIES.

WYRLEY No.3 COLLIERY, Great Wyrley. SE15
Ex W.Harrison Ltd. (SK000067)

WM2 from 1/1/1947; Merged with **MID CANNOCK COLLIERY** 6/1963 & surface CLOSED.

A 2' 0". gauge rope worked overland tubway ran north west from Grove Colliery (which see) to Wyrley No.3 Colliery (1¼ miles). Tubs of 14 cwt capacity ran on this to the washery at Grove. A major reconstruction took place in the late 1950's as part of which locomotives were introduced underground and in the surface stockyard The overland tubway was.either modernised for 2.5 tonne minecars or replaced by a conveyor in 1960.

Stockyard Railway

Gauge: 2ft 6in

4wDM	RH		1944	(a)	(2)
4wDM	RH	441947	1959	(b)	(1)

(a) ex Geo Cohen, Sons & Co Ltd, Kingsbury, Warwicks, c4/1958.
(b) ex Lea Hall Colliery, 22/2/1962

(1) to Lea Hall Colliery, 16/8/1962
(2) to Cannock Central Workshops, by 5/1962; returned /1962. May have gone to Mid Cannock Colliery after 8/1963, otherwise s/s.

Underground haulage

Gauge: 2ft 6in

11/2	4wBEF	EE	1810	1952		
		Bg	3354	1952	(a)	(1)
11/3	4wBEF	EE	2297	1956		
		Bg	3455	1956 New		(1)
	4wBEF	EE	2298	1956		
		Bg	3456	1956 New		(2)

(a) ex Littleton Colliery, before /1963.

(1) to West Cannock No.5 Colliery, 10/1963.
(2) to Littleton Colliery, 9/5/1963.

YEW TREE DRIFT MINE, Essington. SF31
(SJ974048)(Approx.)

Opened by WM2 c1948; CLOSED 4/1950. A licensed mine of this name was being operated by A E Marsh by /1952 and by the Yew Tree Colliery Co. Ltd. from c/1953 to 1963. From then until closure, c/1969, it was operated by the Yew Tree Colliery Co (1963) Ltd. It is probable that this was the NCB mine reopened.

No standard gauge rail connection. Narrow gauge track was laid within the drift but without locomotives.

SECTION 4

CONTRACTORS LOCOMOTIVES

T. A. HAWKINS & SONS, LTD.,
Colliery Proprietors,
L.M. & S RAILWAY WHARF.
BRADFORD PLACE :: WALSALL.
House and Works Coal and Slack
direct from Colliery to Consumer
Agent: J. H. SHINGLER.   Telephone **Walsall 4060**

BRADDOCK AND MATTHEWS
LNWR Dallow Lane Branch and Tutbury Spur, Burton on Trent

Work carried out between 1880 -1881 on a branch railway between Shobnall and Stenson Junction, which was the only part built of a projected line to join the LNWR line at Alrewas.

Gauge: 4ft 8½in

	0-6-0ST	IC	MW	166	1865	(a)	(1)
HULTON	0-4-0ST	OC	MW	577	1875	(b)	(2)

(a) ex John Bradley, Shut End Colliery, Kingswinford, c/1879.
(b) ex Gripper and Bayliss, contrs, Southport, Lancahire, c/1880.

(1) to Braddock and Matthews, Bolton-Kenyon Junction widening contract, Lancashire
(2) to Braddock and Matthews, Downholland and Southport contract, Lancashire

A. BRAITHWAITE AND CO
Contract at Penkridge

Locomotive used during the construction of the LNWR sidings at Penkridge and Littleton Colliery branch from 1899 to 1902. Advertised for sale 26/3/1902.

Gauge 4ft 8½in

	0-6-0ST IC	MW	1070	1888	(a)	(1)

(a) supplied new to Logan & Hemingway and used on the Chester to Shotton Railway construction contract, Cheshire/Flint
(1) to Pauling & Co Ltd, Northolt Junction to High Wycombe Railway contract, Bucks, for GW & GC Joint Committee.

THOMAS BRASSEY
Cannock Mineral Railway ballasting 1859

Gauge: 4ft 8½in

	0-4-0	IC	Maudslay & Field	1838	(a)	(1)

(a) ex LNWR, 79, hire 5/1859. Originally London and Birmingham Railway 'No.79'.
(1) retn to LNWR c/1860.

GRANT LYON EAGRE LTD
Littleton Colliery contract 1979-1981

Locomotive engaged in relaying track on the NCB railway from Penkridge to the colliery.

Gauge: 4ft 8½in

| | | 4wDH | S&H | 7512 | 1972 | (a) | (1) |
| ALFRED HENSHALL | | 0-4-0DM | RH | 313392 | 1952 | (b) | (2) |

(a) ex Daw Mill Colliery contract, Warwicks, between 7/1979 and 8/1979
(b) ex Kingsbury contract, Warwicks, /1980.

(1) to Rawdon Colliery contract, Leics, 10/1979.
(2) to Tyne and Wear Metro contract, Pelaw, Tyne & Wear, 6/1981.

GWR ENGINEERS DEPARTMENT
Oxley to Kingswinford track laying 1919-1924

The GWR completed the contract commenced by Perry & Co (Bow) Ltd (which see) using a departmental locomotive after finding that heavier engines could not work along the contractor's line.

Reference: The Railway to Wombourn - N.Williams.

Gauge: 4ft 8½in

PWM 40 (GALLO)   0-4-0ST   OC   MW   1040   1888   (a)   (1)

(a) ex GWR Swindon c/1919. Previously Fishguard & Rosslare Railways & Harbours Co.

(1) retn to Swindon c/1925. Scr /1927.

GEORGE LAW
Kinver Light Railway Construction, 1899 - 1901

The Kinver Light Railway was a passenger carrying electric tramway of the British Electric Traction Co Ltd group, operated by the Dudley, Stourbridge & District Electric Traction Co Ltd. Construction of the permanent way was let to George Law of Kidderminster who used a Kitson steam tram locomotive on this contract.

Reference: Blackcountry Tramways Vol 1, 1872 -1912; J S Webb.

Gauge: 3ft 6in

              -              0-4-0Tram   K   (a)   (1)

(a) ex Dudley & Stourbridge Steam Tramway Co Ltd, c/1889

(1) to Birmingham & Midland Tramways Ltd, c/1901

HENRY LOVATT
Midland Railway Aldridge to Brownhills Railway construction 1880- 1882.

Gauge: 4ft 8½in

Details of locomotives used on this contract not known. One may have been BP 1148/1872 on hire from Walsall Wood Colliery Co Ltd.

PAULING & CO LTD
Coven Contract SB3

Construction of Royal Ordnance Factory, Coven, (SJ 927058), near Wolverhampton, /1941

Gauge: 4ft 8½in

    THIKA   0-6-0ST   OC   WB   2197   1922   (a)   (1)

(a) ex another contract

(1) to another contract

PERRY AND CO (BOW) LTD
GWR Oxley to Kingswinford Junction construction 1913-1919.   SJ2

Work on this line commenced in 1913 and by the start of the First World War the earthworks for much of the track had been built and a contractors railway was in place for most of the route. The yard and locomotive shed was located at Planks Lane, Wombourn. Construction continued until 1916 and even some permanent way had been put down before hostilities caused most work on the line to cease. A workmen's service was operated along part of the line to convey Perry's workers who were redeployed for work at other locations. The locomotives THEO and GLADYS were used on this service. Many of the locomotives were requisitioned for use at munitions factories elsewhere. After the war the remaining Perry locomotives and plant were disposed of through a sale 5/1920 and the work completed by the GWR. Horse worked narrow gauge tramroads were also employed on this contract.

Ref: The Railway to Wombourn-N.Williams.
     Information collected by late R. T. Russell from Charles Thomas, a driver here, 1911-1915, 1921-1925.

Gauge: 4ft 8½in

BASIL	0-6-0ST	OC	HC			(a)	(1)
ADA	0-6-0ST	IC	MW	458	1874	(b)	(2)
THEO	0-6-0ST	IC	MW	973	1886	(c)	s/s c/1919.
JOHN	0-6-0ST	IC	HC	327	1889	(d)	(3)
DAVID	0-6-0ST	IC	MW	774	1881	(e)	(4)
ALDWYTH	0-6-0ST	IC	MW	865	1883	(f)	(5)

GLADYS	0-6-0ST	IC	MW	1145	1890	(g)	(6)
IRENE	0-6-0ST	IC	MW	1502	1900	(h)	(7)
ARTHUR	0-6-0ST	IC	MW	1601	1903	(i)	(2)

(a) ex ?
(b) previously R.T.Relf & Son, contr, Truro, Cornwall
(c) previously A. Kellett, contr, FRANKLEY.
(d) ex Edmund Nuttall, contractors, DEIGHTON; previously Poynton Collieries, Cheshire.
(e) originally Lucas & Aird, contrs, '237'.
(f) ex ? , by 16/12/1913; originally Lucas & Aird, contrs, '339'.
(g) previously T.Oliver & Son, contr, Rugby, Warwicks
(h) ex Port Talbot contract, Glamorgan
(i) ex John Aird, Avonmouth contract, Somerset '138'; here by 16/12/1913.

(1) to Earl of Dudley, Round Oak Steelworks, Brierley Hill.
(2) to Austin Motor Co Ltd, Longbridge, Warwickshire c/1917.
(3) to Ministry of Munitions, National Filling Factory, Grimston, Banbury, Oxon, c/1917.
(4) to Guy Pitt & Co Ltd, Shut End Colliery, Kingswinsford, c/1919.
(5) to Air Ministry, Kenley, Surrey.
(6) to Wolverhampton Corrugated Iron Co Ltd, Ellesmere Port, Cheshire.
(7) to another contract, /1921; s/s after 8/1922.

The use of ALDWYTH on this contact has not been entirely confirmed. In Charles Thomas's account of the locomotive fleet a locomotive TIGER 0-6-0ST IC is included, but ALDWYTH is not mentioned.

A.STREETER & CO LTD
Sewer Contract, Lichfield.

At least one locomotive used on this contract, the tunnelling having been sub contracted to J.J. Gallagher (London) Ltd.

Gauge : 1ft 6in

JM83		4wBE	CE	5942A	1972	(a)	(1)

(a) ex Welham Plant Ltd, St Neots, Cambs, 7/4/1986.

(1) retn to Welham Plant Ltd, 17/7/1986.

SECTION 5

PRESERVATION SITES

ALL QUOTATIONS SUBJECT TO THE USUAL STRIKE, LOCK-OUT, ACCIDENT, WAGES & TO ALTERATION IN RAILWAY RATES, &c., CLAUSES AND ALL OTHER CONTINGENCIES, AND TO ACCEPTANCE OF SAME BY RETURN OF POST.

All Communications to be addressed to the Firm.

T.A. HAWKINS & SONS
LIMITED.

TELEPHONE, 16 CANNOCK.
TELEGRAMS, HAWKINS COLLIERY, CHESLYN HAY.
PASS. STATION, WYRLEY & CHESLYN HAY, L.&N.W.Rly.

Cannock Old Coppice Colliery,

Cheslyn Hay,

via Walsall.

South Staffordshire Handbook. Page 125

BASS LTD.
BASS MUSEUM, Horninglow Street, Burton on Trent. SH3

The museum is built upon the site of the former Bass workshop at the corner of Guild Street and Horninglow Street (SK 248234). Many aspects of Burton's brewing industry and the former Bass railway are depicted here. Exhibits include two of the locomotive fleet which stand under the canopy of a goods station recreated at the museum. The railway coach preserved here was used by T.A.Walker as a pay/inspection coach whilst his firm engaged in building the Manchester Ship Canal. After its purchase by Bass the coach continued to be used on similar duties.

Gauge: 4ft 8½in

No.9		0-4-0ST	OC	NR	5907	1901	
			Reb	Bass		1956	(a)
No.20		4wDM		KC		1926	(b)

(a) ex Staffordshire County Museum, Shugborough Hall, 4/4/1977.
(b) ex Chasewater Light Railway Co Ltd., Norton Canes, c6/1980, loan.

CHASEWATER LIGHT RAILWAY AND MUSEUM CO LTD
Chasewater Light Railway Co Ltd until /1987
Railway Preservation Society until /1972

This society was inaugurated in October 1959 and is involved with the preservation of locomotives and rolling stock. Having a country-wide interest, the RPS set up three groups around Britain in Scotland, London and the Midlands. The Midland group initially had a depot at the BR Hednesford Goods yard (SK005132). Here locomotives and stock were crammed into the confined sidings or were stored at other temporary locations. In 1966 this group acquired a section of track from the NCB beside the Chasewater Reservoir. This track was essentially part of the Cannock Chase Colliery system, but there was also a short spur which had once connected with Conduit Colliery. Gradually stock was moved to the new site and stored in the open at the southern end of the line. Here a depot and a short platform were built which took the name of Brownhills West (SK035070). By July 1970 the move to Chasewater was complete and BR handed the Hednesford Goods yard over to private development. From 1969 a passenger service was operated over the line on summer Sundays, but this was suspended between 1983 and 1986 whilst the track was relaid and Brownhills platform rebuilt. During 1981 a locomotive shed (SK032072) was erected.

HEDNESFORD DEPOT SC17

Gauge: 4ft 8½in

1054		0-6-2T	IC	Crewe	2979	1888	(a)	(1)
9	CANNOCK WOOD	0-6-0T	IC	Bton		1877	(b)	(2)

(a) ex BR, Crewe, 4/1961.
(b) ex NCB, Rawnsley Depot, 12/1963.

(1)   to National Trust, Penrhyn Castle, Caernarvonshire, 12/1963.
(2)   to Cannock Wood Colliery 25/6/1970, for transfer to Chasewater Depot.

CHASEWATER DEPOT   SE14

Gauge: 4ft 8½in

	CHASEWATER No.1	4wDM		FH	2914	1944	(a)	(1)
	LANCE	0-4-0ST	OC	P	1038	1906	(b)	Scr 3/1972
		0-4-0ST	OC	P	1823	1931	(c)	Scr 3/1972
20		4wDM		KC		1926	(d)	(2)
21		4wDM		FH	1612	1929	(d)	
		0-4-0F	OC	AB	1562	1917	(e)	Scr 3/1973
		0-6-0ST	OC	HC	431	1895	(f)	
4	ASBESTOS	0-4-0ST	OC	HL	2780	1909	(g)	
11	ALFRED PAGET	0-4-0ST	OC	N	2937	1882	(h)	
3		0-4-0ST	OC	AB	1223	1911	(i)	
1		4wPM		MR	1947	1919	(j)	
9	CANNOCK WOOD	0-6-0T	IC	Bton		1877	(k)	(3)
No.2	THE COLONEL (LION)	0-4-0ST	OC	P	1351	1914	(l)	
8	INVICTA	0-4-0ST	OC	AB	2220	1946	(m)	
10	WHIT No.4	0-6-0T	OC	HC	1822	1949	(n)	
6		0-4-0ST	OC	P	917	1902	(o)	
7		0-4-0DE		RH	458641	1963	(p)	
DB 97005		4w-4wDMR		Wkm	7346	1957	(q)	
5		4wVBT	VCG	S	9632	1957	(r)	
37	TOAD	0-4-0DH		JF	4220015	1962	(s)	

(a)   ex Pittsteel Ltd, Aldridge, 4/1966.
(b)   ex Whitecross Co Ltd, Warrington, Lancashire, 10/1966.
(c)   ex Ambrose Shardlow & Co Ltd, Sheffield, 16/06/1967.
(d)   ex Bass, Mitchells & Butlers Ltd, Burton-on-Trent, 7/1967.
(e)   ex W.M.Crawford & Sons Ltd. Liverpool, 11/1967.
(f)   ex Staveley Minerals Ltd, Desborough Quarries,Northants, 2/12/1967.
(g)   presented to society by Turners Asbestos Cement Co Ltd, Trafford Park, Manchester 15/6/1968.
(h)   ex Baird's & Scottish Steel Ltd, Gartsherrie Works, Lanarkshire, per Thos.W.Ward Ltd, 23/6/1968.
(i)   ex N.Greening (Warrington) Ltd, Lancashire, 7/1966. Stored at Hixon, Staffordshire until 9/1968.
(j)   ex Rylands Ltd, Warrington, Lancashire, 14/03/1970. Originally LYR No.1.
(k)   ex Hednesford Depot, via Cannock Wood Colliery, 27/6/1970.
(l)   ex Wallsend Slipway & Engineering Co Ltd., Wallsend-on-Tyne, 23/7/1974 and moved to Chasewater 1/1975.
(m)   purchased by Mike Wood from Stanier Black Five Locomotive Society, Bridgnorth, Shropshire, 11/3/1975 and moved to Chasewater 5/6/1975.
(n)   ex Yorkshire Dales Railway Society, Embsay, North Yorkshire, 18/2/1978.
(o)   presented on long term loan by Albright & Wilson Ltd, Oldbury, 4/8/1978.
(p)   ex NCB Whitwell, Derbyshire, c4/1979 to Manpower Services Commission for use on Chasewater Light Railway; later purchased.
(q)   ex BR, 2/10/1981. Delivered Chasewater 7/10/1981.
(r)   ex Midland Railway Preservation Society, Ripley, Derbyshire, 20/2/1982.

(s) ex Royal Ordnance plc, Radway Green Factory, Alsager, Ches., c6/1987 (after 1/12/1986, by 19/7/1987)

(1) to Brian Roberts, Hill Farm Railway, Tollerton, Nottinghamshire, 4/1986.
(2) presented on long term loan to Bass Museum, Burton on Trent, c6/1980.
(3) to Lord Fisher Loco Group, East Somerset Railway, Cranmore, Somerset, 11/9/1978.

LICHFIELD DISTRICT COUNCIL
BEACON PARK, Lichfield. SG1

Locomotive stood in the play area besides a Fowler steam roller.

Gauge: 4ft 8½in

| | | 4wDM | FH | 3809 | 1963 | (a) | Scr c/1987 |

(a) ex Rom River Reinforcement Co Ltd, Lichfield, 10/1973, presented by Mr Taroni.

J. STRIKE & N. CURTIS
Tamworth

Locomotive preserved at a private location.

Gauge: 2ft 0in

| 87033 | 746 | 4wDM | MR 40SD501 | 1975 | (a) |

(a) ex Severn Trent Water Authority, Tame Division, Minworth, West Midlands, /1990

SECTION 6

ENGINEERS and DEALERS

Burton Constructional Engineering Co. Ltd.
BURTON UPON TRENT

Telephone 2031   Telegrams: "Structural Burton upon Trent."

Makers of ...

STEEL ROOFS
STEEL BUILDINGS : BRIDGE WORK
COLLIERY STRUCTURES
COLUMNS : GIRDERS : CHIMNEYS
AERIAL ROPEWAY STEELWORK
STEELWORK OF RIVETED OR WELDED CONSTRUCTION

Designs and Estimates free on application.

In addition to industrial railways which fill the bulk of this handbook, there are several other locations which deserve mention. Firstly there were several firms which built locomotives. Then there were others which dealt in their resale and undertook repairs. Several contractors were based in the Midlands and their yards held locomotives from time to time. Fourthly there were the scrap merchants whose yards held locomotives for a period of days to years. Finally there is a 'do not know' category which includes locomotives supplied to customers which cannot be matched with any known South Staffordshire location, and locomotives advertised in the press from ill-defined locations .

In no respect can this be a definitive list and details have been kept brief. Yet here below is an insight into less well known side of the industrial railway field.

BAGULEY DREWRY CO LTD
BURTON ON TRENT WORKS   SH18
                                                                  SH19
E.E Baguley until /1967
Baguley Engineers Ltd until /1932
Originally Baguley Cars Ltd

The original works (**SH18**, SK238228) was established in 1903 beside the Midland Railway's Shobnall Branch at Burton by the Ryknield Engine Co Ltd (Ryknield Motor Co Ltd from 4/1906). Standard gauge sidings served the factory by 1904. In 1911 the premises were acquired by Baguley Cars Ltd (registered 3/9/1911). The company built a few steam locomotives, but the main products of the works were petrol and diesel locomotives. Railcars were also made, as were several electric locomotives. In the early years the company built railway vehicles for the Drewry Car Company using the works number series 2000 - 2999, while its own products were in the 3000 series. In 1934 the business was transferred to a new factory in Uxbridge Street (**SH19**, SO243525), with sidings connected to the Bond End Branch. From 1937 all locomotives were allocated numbers in the 3000 series. All production has now ceased.

WILLIAM BRIERS AND SONS (TAMWORTH), Glascote.

A scrapyard which has had locomotives for scrap including:
2ft 0in gauge 4wDM RH 402815 of 1956; ex Baggeridge Brick Co, Kingsbury Brick and Tile works.

HODGES & PORTER, Burton-on-Trent

A builder at Horninglow who received 2ft 0in gauge 4wPM MR 4519 new in 1928. It is not known where it worked.

J.T.LEAVESLEY, Surplus Stores, Alrewas.

Plant stored at a depot beside the A38 at Alrewas included a 2ft gauge 4wPM.

J.MURPHY & SONS LTD, Contractors
Hawks Green Lane, Cannock.

Battery locomotives stored in this yard (SJ994108) between contracts.

THORNEWILL & WARHAM (1919) LTD
NEW STREET WORKS, Burton on Trent.   SH17
Thornewill & Warham Ltd until /1919

This firm constructed items of steam driven plant including winding engines, boilers and railway locomotives. It commenced building locomotives during the late 1850's. A single series of works numbers was used for all products so that the works numbers do not reflect the total number of locomotives built, although this number must be at least thirty. Many of the Burton Breweries used Thornewill & Warham locomotives as did several of the mines in North Staffordshire. At first, the works (SK247227) had no rail connection and all products were taken out by road. About 1881 a rail siding was laid to connect the factory with the Robinsons Brewery Branch. Locomotive building ceased about 1900 though repairs and rebuilding work was continued for several years.

SECTION 7

NON-LOCOMOTIVE LINES

ASK FOR

IND COOPE'S

CELEBRATED

PALE ALES

IN BOTTLE AND ON DRAUGHT.

ALBION BREWERY (BURTON-ON-TRENT) LTD
ALBION BREWERY, Burton-on-Trent   SH11
Mann Crossman & Paulin until 3/1896

Brewery (established 1874) (SK231203) served by Midland Railway Shobnall Wharf branch. The sidings were extended into the brewery 29/12/1876 and traffic worked by the Midland Railway. Brewery acquired in 1902 by Marston, Thompson and Son (see Section 1).

MARQUIS OF ANGLESEY
HAYES COLLIERY, Brereton.   SA3

A tramroad built about 1795 linked the Hayes Colliery (SK046152) with a wharf on the Trent and Mersey Canal. The line, about two miles in length, took a circuitous route which passed through the centre of Rugeley. Wagons of coal from the mine were raised up an incline to Hill Top and then taken along the tramroad by horses. By 1854 the mines had passed to the Earl of Shrewsbury who operated the Brereton Colliery adjacent.

Ref: A Transport History of Cannock Chase - R.Francis.

BIRMINGHAM & LIVERPOOL JUNCTION CANAL

Parts of this canal follow the Staffordshire and Shropshire county boundary. It was built 1826 - 1836 to the designs of Thomas Telford and the works included several embankments and deep cuttings. Telford used several narrow gauge tramways during the construction, including:
1   A tramway laid down by 7/1830 from Mrs Johnsons Wood to Thelmore Embankment.
2   A tramway, in place by 1/1831, which ran north from Wheaton Aston Ridge
3   A tramway, in place by 1/1831, in Brewood Deep Cutting from where wagons carried spoil to form embankments across adjacent valleys.

BROWNHILLS COMMON TRAMROAD

A tramroad about 800 yards long was built to coal pits owned by the Hanbury family on Brownhills Common (SK034062). The line was laid from a basin, at The Slough, on the Wyrley and Essington Canal at the Slough across Coppice Lane to mines near Watling Street. Pits later worked as part of Coppice Colliery which see.

BROWNHILLS TRAMROAD

This tramroad, about 1400 yards in length, ran north to coal pits near Watling Street from a canal wharf on the Wyrley and Essington Canal near Brownhills.

Ref: Stone Blocks and Iron Rails- B.Baxter.

CANNOCK & HUNTINGTON COLLIERY CO LTD
LITTLETON COLLIERY, Huntington   SB4

Sinking of this mine (SO972128) was commenced in December 1872 and a 2ft gauge tramway was laid, between 1877 and 1878, linking the mine with the Staffordshire and Worcestershire canal at Otherton (SO933118). The line was used to convey, from the canal, materials for the sinking of the mine and was horse worked, assisted by rope. Tremendous difficulties were encountered with the sinking of the mine, particularly with water flooding the shaft and eventually sinking operations were suspended. The plant was auctioned in 8/1884 and the company wound up the same year. From 1897 sinking operations were recommenced on the same site, by the Littleton Collieries Ltd (see section 1).

CANNOCK & LEACROFT COLLIERY CO LTD.
LEACROFT COLLIERY, Leacroft, near Cannock   SD30
Registered 12/1871

The sinking of these mines (SJ997096) was commenced in 1874. On completion, they were linked to the Cannock Extension Canal by a narrow gauge tramway c750yds long. This line ran SW to a wharf known as Leacroft Wharf passing through, en route, a tunnel c150 yards in length. In 1879 the LNWR built a new line between Norton and Hednesford which passed close to the colliery. Standard gauge sidings were laid to connect with the LNWR line and wagons were shunted by gravity and capstans. Mine taken over by NCB, WM Division, Area 2, 1/1/1947.

CENTRAL ELECTRICITY GENERATING BOARD
BURTON POWER STATION   SH36
Central Electricity Authority until 1/1/1958
British Electricity Authority until 1/4/1955
Burton Corporation until 1/4/1948

Sidings (SK254244) served this power station and were connected to the Hay Branch via those of the adjacent Gas Works (which see under East Midlands Gas Board). Closed c/1970.

JAMES EADIE
CROSS STREET BREWERY, Burton on Trent　　　　　　　　　　　　　SH32

This brewery, with stores and cooperage, (SK 245229) was not served by rail until the Midland Railway completed its Duke Street branch in 1875. This double track branch ran into Bass' brewery, with a realignment of the adjacent Russell Street, and a siding served Eadie's Brewery.

EAST CANNOCK COLLIERY CO LTD
EAST CANNOCK COLLIERY, Hednesford.　　　　　　　　　　　　　SC15
Registered 2/11/1880

Sinking operations began in June 1871 and coal was reached by 1874. The mine (SJ998112) was connected, in 1877, to the LNWR Walsall-Rugeley line by a short mineral branch. A narrow gauge tramroad conveyed pit tubs down to a wharf at the end of the Cannock Extension Canal, 250 yds distant. Mine taken over by NCB, WM Division, Area 2, 1/1/1947.

EAST MIDLANDS GAS BOARD
BURTON GASWORKS　　　　　　　　　　　　　　　　　　　　　　SH30
Burton Corporation until 1/5/1949

Standard gauge sidings connected with the Midland Railway Hay Branch by 1895. These works were established by Burton Corporation and when the gas industry was nationalised, they passed to the Nottingham and Derby Division of the East Midlands Gas Board.

ESSINGTON FARM COLLIERY CO LTD
ESSINGTON FARM COLLIERY

This company possessed two short tramways which ran to the Wyrley and Essington Canal. Their history is dealt with in section 1.

EVERARD & CO
TRENT BREWERY, Burton on Trent　　　　　　　　　　　　　　　　SH33

Sidings connected this brewery (SK 241227) with the Midland Railway Bond End branch.

FERRO (GREAT BRITAIN) LTD
WOMBOURN WORKS, Wombourn, near Wolverhampton. SJ30
prev Ferro Enamels Ltd

This firm of enamel manufacturers, had a 2ft gauge hand worked tramway, laid within the works, used for carrying materials around the shop floor.

MATTHEW FROST, Contractor

Matthew Frost of Bilston was responsible for the following contracts:
    1839-1841 Construction of the Staffordshire and Worcestershire Canal Company, Hatherton Branch and reservoir.
    1841-1842 Cheslyn Hay Tramroad.

For other contracts relating to Matthew Frost see West Midlands Handbook

J.GIBB AND SONS
LNWR NORTON EXTENSION RAILWAY CONSTRUCTION.

Mr Gibb completed the construction of the freight line from Norton Green Colliery to Hednesford in 1878.

BERNARD GILPIN
WYRLEY COLLIERY, Cheslyn Hay. SD31

The Churchbridge Edge Tool Works (SJ986082) was built, in 1806, beside Watling Street for William Gilpin (who see). A narrow gauge tramway was laid c1817 running N from coal pits at Wyrley and Landywood to the Churchbridge Works and a landsale wharf beside Watling Street. Coal and some ironstone was worked from the Wyrley seams. Proprietorship of the tramway and mines changed several times within the Gilpin family. By 1838 (Tithe Survey) William Lawrence Gilpin had charge of the tramway and the Churchbridge works. There was a branch to serve Pennyfield Colliery (SJ983073) while another ran to coal pits and a brickworks south of Cheslyn Hay village. The line terminated on land owned by Rugeley School (SJ986068), where there were several other shafts. In 1838 the Staffordshire & Worcestershire Canal Company proposed to build a branch canal to Churchbridge and in April 1841 attempted to reach an agreement with the members of the Gilpin family for the use of their tramroad to reach the mines at Cheslyn Hay. Negotiations were hampered by Frederic Gilpin and in November 1841 the Canal Company instructed its contractor Matthew Frost to proceed with a separate line which became known as the Cheslyn Hay Tramroad (which see). By 1848 Bernard Gilpin, younger brother of William Lawrence, was in charge of the tramway, which at this time served three mines, Brownsfield, New Wyrley and Rugeley School, all south west of Cheslyn Hay. Bernard also leased from Lord Hatherton land near Fishers Farm at Cheslyn Hay close to the end of the Canal Tramroad. In March 1854 Bernard Gilpin was granted the right to carry coal along the Cheslyn Hay Tramroad from his 'Hatherton Colliery', but not from his other mines. Further developments came with the building of the BCN Wyrley Bank Canal. In December 1864 more land was leased from Lord Hatherton at Fishers Farm in order to lay a tramroad from the Wyrley Mines to the Wyrley Bank Canal where a basin (Gilpins Basin, SJ977058) was built to handle his traffic.

By 1875 three pits and a brickworks remained, the pits being collectively known as the Churchbridge Colliery.. Most of the Wyrley coal seams were worked out, but a new company, the Great Wyrley Colliery Co Ltd (which see) was formed to prove and work deeper coal. Bernard Gilpin sold his existing plant and tramways to the new company. See section 1 for later history.

WILLIAM GILPIN Senior & CO LTD
William Gilpin Senior until 18/4/1881

The Churchbridge Edge Tool Works received its supply of coal over the tramway mentioned above. Track was laid around the works and onto a wharf beside the Staffordshire and Worcester Canal. Management of this factory, and the associated Wedges Mill, was carried out by members of the Gilpin family. Eventually the job fell to Frederic Henry Gilpin who also worked part of the Wyrley Colliery in his own name. When Frederic retired from business (c1874) Churchbridge Works was handed over to Ernest and Charles Burnett who continued to trade under the title of William Gilpin Senior and Co. Standard gauge sidings also served the factory by 1895, and were still isted in 1956. These joined the LNWR/LMS Churchbridge branch near the railway and canal interchange basin. Tramways removed after 1924.

WILLIAM HARRISON LTD

Details of their mining operations at Mid Cannock and Wyrley No.3 collieries are dealt with in section 1. See also West Midlands Handbook.

HENRY HAWKINS
LONGHOUSE COLLIERY AND BRICKWORKS, Cannock. SD32
Joseph Palmer (until 2/1875)

Tramway laid at brickworks (SJ976083) and adjacent colliery. By 1865, and until c1876, the line crossed the Staffordshire & Worcestershire canal to a wharf and canal basin at Long House Farm. Closed by 1886.

HILTON COLLIERY Essington.

Tramway ran SE from pits (SJ974028) to a basin beside the Wyrley and Essington Canal. Mines worked by Essington Farm Colliery Co. (which see) by 1894.

HOCKLEY HALL COLLIERY CO LTD
HOCKLEY HALL COLLIERY & CHEMICAL WORKS   SK30
James Evers-Swindell (until /1872 ?)
Birmingham Banking Co by 12/1864
W H Beaumont until at least 7/1864

Hockley Hall Colliery was originally known as Kingsbury Wood Colliery. In 7/1864 W H Beaumont gave notice to the Midland Railway that the Kingsbury Wood Colliery intended to work the broach coal under the railway. In 12/1864 the Birmingham Banking Co applied to the Midland Railway to lay a siding to the new pits at Kingsbury Wood and in 1/1865 the railway company approved the construction of this siding which connected with their line about a mile south of Wilnecote Station.
The mine was later purchased by James Evers-Swindell who also owned the Homer Hill Colliery near Cradley (which see in West Midlands handbook). Later the colliery was operated by the Hockley Hall Colliery Co Ltd (registered 1872). In 1878 the business was merged with the Whateley Colliery Co Ltd to form the Hockley Hall & Whateley Collieries & Brickworks Ltd. (which see in Warwickshire Handbook for subsequent history).

IND COOPE LTD
SHOBNALL RD MALTINGS and CURZON STREET BOTTLING STORES
                                                    SH39
Ind Coope & Allsopp Ltd until 31/12/1958
Ind Coope Ltd until 6/1934

Extensive premises (SK240231) served by a siding which ran west from Burton Station and which originally formed part of the through route to Shobnall until the Bond End Branch opened. Rail traffic ceased in the 1960's and premises closed by 1980.

McKENZIE, STEPHENSON & BRASSEY

Contractors employed on the construction of the Rugby to Stafford railway (the Trent Valley line, 50 miles), where a large number of men and horses were engaged on the work.

MID CANNOCK COLLIERY CO LTD

See entry under William Harrison Ltd, (Section 1).

MIDLAND JOINERY WORKS LTD
JOINERY WORKS, Lichfield St, Burton on Trent SH38
f C Perks & Sons

Served by a siding from the Bond End Wharf. Narrow gauge hand worked lines were used for internal timber transport.

E. J. MILLER & CO LTD
CRESCENT BREWERY, Burton on Trent. SH37
Thomas Cooper & Co until 7/1919

By 1895 sidings linked the brewery (SK243243), established 1865, with the Midland Railway Horninglow branch. Brewery taken over by Thos. Salt & Co Ltd in 1919 and later closed.

FRANCIS PIGGOT

Contractor for the Cannock and Norton branches of the South Staffordshire Railway, 1855 - 1858. On completion of the civil engineering work for the embankments and cuttings in 1857 he received the contract to lay the permanent way.

QUINTON COLLIERY CO LTD
QUINTON COLLIERY, Great Wyrley. SD33
S. Blewitt & Co Ltd by /1905
H.Blewitt.
E.Sayer.

This mine (SJ989067) takes its name from the Quinton family which owned the land. The shafts were sunk before 1838 and worked by Edward Sayer and others on a small scale. New plant laid down c1903 and standard gauge sidings were built to link mine with the LNWR Walsall to Rugeley line south of Churchbridge station. Mr Blewitt on two occasions tried to sell the enterprise to Patent Shaft & Axletree Co Ltd without success.
Mines closed and plant offered for sale by auction 1908.

ROBINSON'S BREWERY LTD
UNION ST BREWERY, Burton on Trent.   SH31
Thomas Robinson and Co until 3/1896

The brewery, cooperage and maltings (SK 248228) were established in 1863 beside Union Street. Sidings were completed in April 1880 which connected with the Midland Railway New Street branch, which also served the breweries of Bindley & Co, Marston Thompson & Evershed, Robinson's and Worthington's. The business was taken over by Ind Coope Ltd in 1920 and the brewery closed c/1930. The company went into voluntary liquidation 24/4/1929.

GEORGE SKEY & CO LTD

The Peel and Wilnecote collieries worked by G.Skey are discussed in Section 1

SOUTH STAFFORDSHIRE WATERWORKS LTD

This public utility company owned two pumping stations at Lichfield with rail connections. Sandfield, built in 1858, had a siding alongside the LNWR Lichfield to Walsall line about two miles east of Lichfield. Trent Valley, built 1900, was served by a siding at Lichfield (Trent Valley).

STAFFORDSHIRE AND WORCESTERSHIRE CANAL
CHESLYN HAY TRAMROAD.   SD34

Line built in 1842 from a basin (SJ967089) of the Hatherton branch of the South Staffordshire Canal, at Long House Farm, to coal mines at Cheslyn Hay where the canal met the tramway operated by Bernard Gilpin. Coal carried along the line included that from Edward Sayer's, Old Falls Colliery and Bernard Gilpin's Hatherton pit. A branch railway was proposed to connect the tramroad with the South Staffordshire Railway Cannock line but this was never built. By 1880 the tramroad was in a poor state and in March of that year the Great Wyrley Colliery Co Ltd offered to repair the line. From 1881 the bottom (northern) part of the line was worked by this company while the upper end at Cheslyn Hay ceased to be used and was lifted.

JOHN STANLEY

The 1838 Tithe Survey for Great Wyrley mentions a disused tramway and pits near Bradley Nook on the road from Walsall to Great Wyrley in the occupation of John Stanley. Colonel Vernon was the landowner.

TAME VALLEY COLLIERY CO
TAME VALLEY COLLIERY & BRICKWORKS SK7
Woods & Greenwood until c/1864
Greenwood & Co until c/1862
Greenwood & Sinclair until c/1861

Standard gauge sidings were laid in 1860 to connect this mine with the Midland Railway Birmingham to Derby line south of Wilnecote Station. In addition to coal mining the company made red and blue bricks. The concern was acquired by George Skey c/1869.

H. VERNON
ESSINGTON COLLIERY SF30

The collieries at Essington Wood are shown on William Yates map of Staffordshire, which was compiled between 1769 and 1775. A tramway was built in 1799 from these mines (SJ 971035) to the terminus of the Lord Hays Canal at Newtown, Bloxwich. There were several pits here which were managed on behalf of successive generations of the Vernon family by colliery agents. Track lifted c/1855 anfter mines leased by Samuel Mills. (See entry for Darlaston Coal & Iron Co Ltd in Section 1).

PETER WALKER & SON

SHOBNALL BREWERY, Burton on Trent SH40

Standard gauge sidings connected this brewery (SK 237231) with the Midland Railway Shobnall branch south of Wellington Street.

CLARENCE BREWERY, Burton on Trent SH34

This brewery (SK 242226) in Clarence Street was served by sidings which connected with the Midland Railway Bond End branch.

WOLVERHAMPTON & CANNOCK CHASE LIGHT RAILWAY
Wolverhampton & Cannock Chase Railway until /1912

The Wolverhampton & Cannock Chase Railway was incorporated by Act of Parliament in 1901. The line was to link Brownhills No.3, Hollybank and Ashmore Park Collieries with the GWR at Cannock Raod Junction, Wolverhampton. The western part followed a route close to that proposed for the Cannock Chase & Wolverhampton Railway (which see). The scheme was changed to a light railway under a proposal of 1907. This new proposal was sanctioned under the Wolverhampton & Cannock Chase Light Railway Transfer & Amendment Order of 1912, which took effect in 1913. Parts of the earlier scheme were abandoned including the direct line from Hollybank to Brownhills No.3 Colliery. In its place a new line was to be built to join Hollybank Colliery with the LNWR near Cannock Lodge. Running powere were then to be exercised over the LNWR to a new branch built to Brownhills No.3 Colliery.

Little was actually built. A section from Hollybank to Blackhalves Lane was finally constructed 1922 -1924 as part of the railway link to Hilton Main Colliery. It was owned by the LMSR and later BR but was worked exclusively by colliery locomotives.

WYRLEY CANNOCK COLLIERY CO LTD
WYRLEY CANNOCK COLLIERIES (No.1 and 8).   SD35
Wyrley Cannock Colliery Co until /1879

Several shafts (SJ974066, 975057) were sunk by 1865 beside and near the terminus of the Wyrley Bank Canal. During 1880 a tramway, which passed under the Great Wyrley Colliery Company's tramroad to Gilpins Basin, was built running in a westerly direction from the No.8 pit (SJ979062) to the canal . All mines closed c/1884, the plant being offered for sale 10/1884. No.1 mine was later restarted as the Nook Colliery by the Great Wyrley Colliery Co Ltd, (which see).

Telegrams: "BAGULEY, BURTON-ON-TRENT."
Telephone No. **572**.

BAGULEY CARS Ltd.

AND

McEWAN, PRATT & CO.

LIMITED,

ENGINEERS

MAKERS OF

LOCOMOTIVES

Steam and Internal Combustion
from **2** to **20** Tons,

NARROW GAUGE ROLLING STOCK,

CARRIAGES and WAGONS

MOTOR LORRIES

BURTON-ON-TRENT

LONDON OFFICE:
RIVER PLATE HOUSE, 13, SOUTH PLACE, E.C. 2.

INDEXES

ABBREVIATIONS for BUILDERS
INDEX of LOCOMOTIVES
INDEX of OWNERS
INDEX of LOCATIONS

ABBREVIATIONS USED FOR LOCOMOTIVE BUILDERS

AB	Andrew Barclay, Sons & Co Ltd, Caledonia Works, Kilmarnock, Ayrshire.
AE	Avonside Engine Co Ltd., Bristol.
BD	Baguley-Drewry Ltd, Burton on Trent, Staffordshire.
BE	Brush Electrical Engineering Co Ltd, Loughborough, Leicestershire.
BEV	British Electric Vehicles Ltd, Southport, Lancashire,
Bg	E.E. Baguley Ltd, Burton on Trent, Staffordshire (later BD)
BGB	Becorit (Mining) Ltd, Mansfield Woodhouse, Notts (Power pack supplied by HE).
BgC	Baguley Cars Ltd, Burton on Trent, Staffordshire (later BgE).
BgE	Baguley Engineers Ltd, Burton on Trent, Staffordshire (later Bg).
BH	Black, Hawthorn & Co Ltd, Gateshead.
BP	Beyer Peacock & Co Ltd, Gorton, Lancashire.
Bton	London Brighton & South Coast Railway, Brighton Works, Sussex.
Bury	Edward Bury & Co, Clarence Foundry, Liverpool.
Butt	Butterley Company, Ripley, Derbyshire.
CE	International Combustion Ltd, Hatton, Derbyshire (formerly Clayton Equipment Co Ltd)
Chasetown	Cannock Chase Colliery Co Ltd, Chasetown Workshops, Staffs.
Crewe	Crewe Locomotive Works, London & North Western Railway, LMSR, BR.
DeW	De Winton & Co, Union Works, Caernarvon.
EE	English Electric Co Ltd., Dick Kerr Works, Preston, Lancs
EEV	English Electric Co Ltd, Vulcan Works, Newton le Willows, Lancashire (later GECT)
FH	F.C Hibberd & Co Ltd, Park Royal, London (Ripley from 1963).
FW	Fox, Walker & Co Ltd, Atlas Engine Works, Bristol.
GB	Greenbat Ltd, Albion Works, Leeds (formerly Greenwood & Batley Ltd).
GE	George England and Co, Hatcham Ironworks, New Cross, London.
GECT	GEC Traction Ltd, Vulcan Works, Newton le Willows, Lancashire.
Grange	Grange Iron Co Ltd, Belmont, Co Durham.
HC	Hudswell, Clarke & Co Ltd, Railway Foundry, Leeds.
HCR	Hudswell Clarke & Rodgers, Railway Foundry, Leeds (later HC).
HE	Hunslet Engine Co Ltd, Hunslet Engine Works, Leeds.
Heath	Robert Heath & Sons Ltd, Biddulph, Stoke-on-Trent.
HH	Henry Hughes & Co, Loughborough.
HL	R & W Hawthorn, Leslie & Co Ltd, Forth Bank Works, Newcastle upon Tyne.
H(L)	Hawthorns & Co Ltd, Leith Engine Works, Leith near Edinburgh.
JF	John Fowler & Co (Leeds) Ltd, Hunslet, Leeds.
JS	John Smith, Village Foundry, Coven, nr Wolverhampton, Staffordshire.

K	Kitson & Co Ltd, Airedale Foundry, Leeds.
Kay	James C Kay & Co Ltd, Phoenix Foundry, Bury, Lancashire.
KC	Kent Construction & Engineering Co Ltd, Ashford, Kent.
KS	Kerr, Stuart & Co Ltd, California Works, Stoke on Trent, Staffordshire.
L	R.A.Lister & Co Ltd, Dursley, Gloucestershire.
Lill	Lilleshall Co Ltd, Oakengates, Shropshire.
Maudslay & Field	Maudslay,Sons & Field, Lambeth Marsh, London.
Moyse	Locotracteurs Gaston Moyse, La Courneuve, Seine, France.
MR	Motor Rail Ltd, Simplex Works, Bedford (later SMH).
MW	Manning Wardle & Co Ltd, Boyne Engine Works, Hunslet, Leeds.
N	Neilson & Co, Hyde Park Works, Springburn, Glasgow.
NR	Neilson, Reid & Co, Hyde Park Works, Springburn, Glasgow.
NB	North British Locomotive Co Ltd, Glasgow.
NBH	North British Locomotive Co Ltd, Hyde Park Works, Springburn, Glasgow
NBQ	North British Locomotive Co Ltd, Queens Park Works, Polmadie, Glasgow
OK	Orenstein u. Koppel A.G., Berlin-Drewitz, Germany.
P	Peckett & Sons Ltd, Atlas Locomotive Works, St George, Bristol.
Rawnsley	Cannock & Rugeley Colliery Co Ltd, Rawnsley Workshops, Staffordshire.
RH	Ruston & Hornsby Ltd, Lincoln.
RR	Rolls Royce Ltd, Sentinel Works, Shrewsbury, Shropshire (originally Sentinel).
RS	Robert Stephenson & Co Ltd, Newcastle upon Tyne (Darlington after /1901)
RSH	Robert Stephenson & Hawthorns Ltd.
RSHD	Robert Stephenson & Hawthorns Ltd, Darlington Works.
RSHN	Robert Stephenson & Hawthorns Ltd, Newcastle Works.
RWH	R & W Hawthorn Ltd, Forth Banks Locomotive works, Newcastle (later HL).
S	Sentinel (Shrewsbury) Ltd, Battlefield, Shrewsbury.
SS	Sharp Stewart & Co Ltd, Atlas Works, Manchester (until 1888) and Atlas Works, Glasgow (from 1883).
TH	Thomas Hill (Rotherham) Ltd, Vanguard Works, Kilnhurst, South Yorkshire.
TW	Thornewill & Warham, Burton on Trent, Staffordshire.
VF	Vulcan Foundry Ltd, Newton le Willows, Lancashire (later EEV).
WB	W.G.Bagnall Ltd, Castle Engine Works, Stafford.
Wkm	D.Wickham & Co Ltd, Ware, Hertfordshire.
YE	Yorkshire Engine Co Ltd, Meadow Hall Works, Sheffield.

INDEX OF LOCOMOTIVES

AVONSIDE ENGINE CO LTD, Bristol   AE

Steam locomotives

Works No	Date Ex-works	Gauge	Wheel Arrgt	Cylinders	Wheel Dia	Class	See Page
1386	1897	4ft8½in	0-4-0ST	OC 14 x 20	3ft3in	SS	38
1742	1916	4ft8½in	0-6-0ST	OC 14 x 20	3ft3in	B3	79

W G BAGNALL LTD, CASTLE ENGINE WORKS, Stafford.   WB

Steam locomotives

Works No	Date	Gauge	Wheel Arrgt	Cylinders	Wheel Dia	See Page
1122	10.1889	2ft0in	0-4-0IST	OC 5½ x 8¼	2ft0in	62
1429	24.10.1894	1ft8in	0-4-0ST	OC 6 x 9	1ft6in	58
2108	24.10.1923	4ft8½in	0-4-0ST	OC 14 x 20	3ft6½in	42,83
2193	27.11.1922	4ft8½in	0-6-0ST	OC 17 x 24	3ft9in	82,90,96,114,115
2197	.12.1922	4ft8½in	0-6-0ST	OC 13 x 18	2ft9¼in	122
2508	20.4.1934	4ft8½in	0-6-0ST	OC 15 x 22	3ft4½in	61,88
2815	2.3.1945	4ft8½in	0-4-0ST	OC 14 x 22	3ft6½in	42,83
3077	17.6.1955	4ft8½in	0-6-0ST	OC 16 x 24	3ft6½in	96

Diesel locomotives

Works No	Date Ex-works	Gauge	Wheel Arrgt	H.P.	Wheel Dia	Engine	Class	See Page
3117	27.5.1957	4ft8½in	0-6-0DM	208	3ft4in	National	AA5	101,113,114
3118	15.7.1957	4ft8½in	0-6-0DM	208	3ft4in	National	AA5	101,103,107,114
3122	4.11.1957	4ft8½in	0-6-0DM	308	3ft4in	National	AA7	101,103
3123	6.12.1957	4ft8½in	0-6-0DM	308	3ft4in	National	AA7	101.108,115

BAGULEY CARS LTD, Burton on Trent   BgC

Steam locomotive

2001	23.2.1920	4ft8½in	0-4-0ST	OC 13 x 18				77

BAGULEY ENGINEERS LTD., Burton on Trent   BgE

Petrol locomotives

1654	09.1.1928	4ft8½in	0-4-0PM	45	Baguley	44,73
2071	.10.1931	4ft8½in	4wPM	25	Ford	79

E.E.BAGULEY LTD., Burton on Trent Bg

Battery and Diesel locomotives

3027	10.12.1939	4ft 8½in	0-4-0DM	85	3ft 1in	Gardner 4L3	42
3227	14.3.1951	4ft 8½in	0-4-0DM	150		Paxman	68
3354	1.9.1952	2ft 6in	4wBEF	64		EM1A2	106,110,117,118
3357	.1952	4ft 8½in	0-4-0DM	150		Paxman	68
3410	7.2.1955	4ft 8½in	0-4-0DM				73
3454	4.1.1956	2ft 6in	4wBEF	64	2ft 0in	EM1A2	110
3455	13.7.1956	2ft 6in	4wBEF	64	2ft 0in	EM1A2	117,118
3456	14.7.1956	2ft 6in	4wBEF	64	2ft 0in	EM1A2	110,118
3509	10.9.1958	4ft 8½in	0-4-0DM	204		Gardner 8L3	42
3547	.1960	2ft 6in	4wBEF	64		EM4A1	117
3554	.1961	2ft 6in	4wBEF	64		BML1	106
3568	23.5.1961	4ft 8½in	0-4-0DM	204		Gardner 8L3	42
3582	17.3.1962	2ft 6in	4wBEF	64		BM41	110
3589	9.11.1962	4ft 8½in	0-4-0DM				42
3590	11.12.1962	4ft 8½in	0-4-0DM				42

ANDREW BARCLAY, SONS & CO LTD, CALEDONIA WORKS,
Kilmarnock. AB

Steam locomotives

1083	16.12.1908	4ft 8½in	0-4-0ST	OC	16 x 24	3ft 7in	45
1115	29.11.1909	4ft 8½in	0-4-0ST	OC	16 x 24	3ft 7in	45
1223	14.3.1911	4ft 8½in	0-4-0ST	OC	10 x 18	3ft 0in	127
1365	9.6.1914	4ft 8½in	0-4-0ST	OC	14 x 22	3ft 5in	45,89,95
1562	1.9.1917	4ft 8½in	0-4-0F	OC	15 x 18	3ft 0in	127
1576	1.4.1918	4ft 8½in	0-6-0ST	OC	12 x 20	3ft 2in	38
1858	9.2.1925	4ft 8½in	0-4-0ST	OC	14 x 22	3ft 5in	39,68
1944	30.1.1928	4ft 8½in	0-4-0F	OC	15 x 18	3ft 0in	56
2220	16.10.1946	4ft 8½in	0-4-0ST	OC	12 x 20	3ft 2in	127
2247	26.1.1948	4ft 8½in	0-4-0ST	OC	16 x 24	3ft 7in	93.98.100.112

Diesel locomotives

342	13.09.1940	4ft 8½in	0-4-0DM	153	Gardner 6L3	73
344	21.01.1941	4ft 8½in	0-4-0DM	153	Gardner 6L3	73
357	29.09.1941	4ft 8½in	0-4-0DM	153	Gardner 6L3	73,80

BECORIT (MINING) LTD, Mansfield Woodhouse, Notts BGB

25/2/216 .1971 trapped rail system 200m 1 axle drive DHF 97.111
 25hp fitted with Perkins 3152 engine

BEYER, PEACOCK & CO LTD., Gorton, Manchester   BP

Steam locomotives

28	.1856	4ft8½in	0-4-2ST	IC	14 x 20	4ft0in	49.93.98
204	11.2.1861	4ft8½in	0-4-2ST	IC	14 x 20	4ft0in	49,93
462	20.6.1864	4ft8½in	0-4-2ST	IC	14 x 20	4ft0in	49,93
794	22.4.1867	4ft8½in	0-4-2ST	IC	14 x 20	4ft0in	49
1140	14.12.1871	4ft8½in	0-4-2ST	IC	14 x 20	4ft0in	79,82,115
1148	9.7.1872	4ft8½in	0-4-0ST	OC			121
1211	14.8.1872	4ft8½in	0-4-2ST	IC	14 x 20	4ft0in	49.93
1915	22.12.1879	4ft8½in	0-4-2ST	IC			52,53

BLACK, HAWTHORN & CO LTD, Gateshead.   BH

Steam locomotives

| 363 | 16.3.1876 | 4ft8½in | 0-4-0ST | OC | 12 x 19 | 3ft2in | 07.1875 | 53 |

BRUSH ELECTRICAL ENGINEERING CO LTD, Loughborough   BE

Steam locomotive

| 317 | .1909 | 4ft8½in | 0-4-0ST | OC | 61 |

Electric locomotives

| 16329 | c/1917 | 2ft6in | 4wWE | 60 |
| 16330 | c/1917 | 2ft6in | 4wWE | 60 |

BUTTERLEY COMPANY, Ripley, Derbyshire   Butt

Steam locomotives

| | .1889 | 4ft8½in | 0-4-0ST | OC | 45,89 |

EDWARD BURY AND CO, CLARENCE FOUNDRY, Liverpool   Bury

Steam locomotives

| | | 4ft8½in | 0-4-0 | IC | 14 x 20 | 3 0 | 52 |

CANNOCK CHASE COLLIERY CO LTD, CHASETOWN WORKSHOPS (CCWR)   Chasetown

Steam locomotive

| | .1946 | 4ft8½in | 0-4-2ST | IC | 14 x 20 | 4 0 | 49,89,93,98 |

CANNOCK & RUGELEY COLLIERY CO LTD, Rawnsley   Rawnsley

Steam locomotive

| | .1888 | 4ft8½in | 2-4-0T | OC | 15 x 20 | 3 6 | 52,89,95 |

South Staffordshire Handbook. Page 150

CLAYTON EQUIPMENT CO LTD., Hatton, Derbyshire   CE

Battery locomotives

4727	.1964	2ft6in	4wBEF	1½ Ton	5	105
4805	.1964	2ft6in	4wBEF	2 Ton	5	105
4960	.1964	2ft6in	4wBEF	1½ Ton	5	104,105
5074	.1966	2ft6in	4wBEF	2 Ton	5	97,104,105,117
5097	.1966	2ft6in	4wBEF	2 Ton	5	104,106
5896A	.1972	2ft6in	4wBEF	CRT3½	17½	106
5896B	.1972	2ft6in	4wBEF	CRT3½	17½	106
5942A	.1972	1ft6in	4wBE	1.75 Ton	7	123
5962A	.1973	2ft6in	4wBEF	CRT3½	17½	106
5962B	.1973	2ft6in	4wBEF	CRT3½	17½	106
B0909A	30.3.1976	2ft6in	4wBEF	CRT3½	17½	106
B0909B	30.3.1976	2ft6in	4wBEF	CRT3½	17½	106
B1828A	10.1.1979	2ft6in	4wBEF	CRT3½	17½	110
B1828B	26.1.1979	2ft6in	4wBEF	CRT3½	17½	106,110
B1886A	22.1.1980	2ft6in	4wBEF	CRT3½	17½	110
B1886B	23.1.1980	2ft6in	4wBEF	CRT3½	17½	106
B1894	4.3.1980	2ft6in	4wBEF	CRT3½	17½	110
B2927	19.8.1981	2ft6in	4wBEF	CRT3½	17½	110
B3428	24.3.1988	2ft6in	4wBEF	CEB4	17½	110

DE WINTON & CO, UNION WORKS, Caernarvon.   DeW

Steam locomotive

        c/1875   2ft0in   0-4-0VBT   62

GEORGE ENGLAND & CO, HATCHAM IRONWORKS, London   GE

Steam locomotive

        4ft8½in   2-2-0ST   3C   52

ENGLISH ELECTRIC CO LTD, DICK KERR WORKS, Preston.   EE

Battery locomotive

533	.1922	4ft8½in	4wBE 38HP	2ft9in	7ft6in	15T	Type 3B	68

The following EE numbers are allocated to locomotives built at other works (which see):

1810	(See Bg 3354)	106,110,117,118
2296	(See Bg 3454)	110
2297	(See Bg 3455)	117,118
2298	(See Bg 3456)	110,118
2660	(See RSHN 7945)	110
2741	(See RSHN 8129)	110
2842	(See Bg 3547)	117
2844	(See RSHN 8131)	97,110

2845	(See RSHN 8132)						97,117
2846	(See RSHN 8133)						97,105
2847	(See RSHN 8134)						97,117
2860	(See RSHD 8203)						97,110
2861	(See RSHD 8204)						105
2862	(See RSHD 8205)						110
2927	(See RSHD 8206)						110
3146	(See RSHD 8288)						105
3147	(See RSHD 8289)						105
3151	(See RSHD 8291)						106,110
3152	(See RSHD 8292)						110
3156	(See RSHD 8296)						116
3157	(See RSHD 8297)						110,116
3161	(See RSHD 8302)						116
3223	(See RSHD 8344)						105
3224	(See RSHD 8345)						105
3400	(See RSHD 8420)						105
3402	(See RSHD 8421)						106,116

ENGLISH ELECTRIC CO LTD, VULCAN WORKS, Newton le Willows EEV

Diesel and Battery locomotives

D1120	14.11.1966	4ft 8½in	0-6-0DH	380	Dorman 6QT	96,103,108
3493	.1964	2ft 6in	4wBEF	64	EM2B1	105
3494	.1964	2ft 6in	4wBEF	64	EM2B1	105
3495	.1964	2ft 6in	4wBEF	64	EM2B1	117
3652	17.3.1965	2ft 6in	4wBEF	64	EM2B1	110
3768	.1966	2ft 6in	4wBEF	64	EM2B1	102,106
3769	1.4.1966	2ft 6in	4wBEF	64	EM2B1	110
3840	.1970	2ft 6in	4wBEF	64	EM2B1	106,117
3841	.1970	2ft 6in	4wBEF	64	EM2B1	102,110
3995	7.5.1971	2ft 6in	4wBEF	64	EM2B1	106

JOHN FOWLER & CO (LEEDS) LTD, Hunslet, Leeds JF

Steam locomotive

1573	1.1873	4ft 8½in	0-4-0ST	OC	(8½ x 14)?	77

Petrol and Diesel locomotives

22000	7.1937	4ft 8½in	0-4-0DM	40	Fowler 4B	80
22982	8.1942	4ft 8½in	0-4-0DM	150	Fowler 4C	80
22989	10.1942	4ft 8½in	0-4-0DM	150	Fowler 4C	80
4220015	4.1962	4ft 8½in	0-4-0DH	203	Leyland	127

FOX, WALKER & CO, ATLAS ENGINE WORKS, Bristol — FW

Steam locomotives

247	6.1874	4ft8½in	0-6-0ST	OC	13 x 20		Class B	57
266	c3.1875	4ft8½in	0-6-0ST	OC	13 x 20		Class B	52,95,103
286	11.1875	4ft8½in	0-6-0ST	OC	13 x 20		Class B1	57
318	c.1876	4ft8½in	0-6-0ST	OC	13 x 20		Class B1	52
370	.1878	4ft8½in	0-6-0ST	OC	14 x 20	3ft7in	Class B1	74
382	.1878	4ft8½in	0-6-0ST	OC	14 x 20	3ft7in	Class B1	59

GEC TRACTION LTD, VULCAN WORKS, Newton le Willows, Lancs — GECT

Battery and Diesel locomotives

5419	.1977	2ft6in	4wBEF	64	2ft0in	EM2B1	106,117
5420	.1977	2ft6in	4wBEF	64	2ft0in	EM2B1	117
5421	17.5.1977	4ft8½in	6wDE	750			108
5422	17.5.1977	4ft8½in	6wDE	750			108
5433	3.1977	2ft6in	4wBEF	64	2ft0in	EM2B1	117
5468	24.8.1977	4ft8½in	6wDE	750		Dorman 12QT	108
5478	3.11.1978	4ft8½in	6wDE	750		Dorman 12QT	103
5479	.1979	4ft8½in	6wDE	750		Dorman 12QT	108
5480	.1979	4ft8½in	6wDE	750		Dorman 12QT	108
5571	.1978	2ft6in	4wBEF	64	2ft0in	EM2B1	106
5572	.1978	2ft6in	4wBEF	64	2ft0in	EM2B1	106

GRANGE IRON CO LTD., Belmont, Durham — Grange

Compressed Air locomotive

	.1883	2ft6in	0-4-0CA	4½ x 8	53

GREENWOOD & BATLEY LTD, ALBION WORKS, Leeds — GB

Battery locomotives

2745	24.10.1956	2ft6in	4wBEF 2 x15	6T		110
2783	21.2.1958	2ft6in	4wBEF 2 x15	6T		102,105,110
2784	31.12.1958	2ft6in	4wBEF 2 x15	6T		102,105,110
2785	31.12.1958	2ft6in	4wBEF 2 x15	6T		102,105,110
2786	3.3.1959	2ft6in	4wBEF 2 x15	6T		105
2901	5.5.1959	2ft6in	4wBEF 2 x15	6T		105,110,116
2987	18.1.1960	2ft6in	4wBEF 2 x15	6T Type GB6		105,110
2988	18.1.1960	2ft6in	4wBEF 2 x15	6T Type GB6		110
6076	13.11.1962	2ft6in	4wBEF 2 x15	6T Type GB6		116
6090	12.7.1963	2ft6in	4wBEF 2 x15	6T Type GB6		97,104,105
6091	12.7.1963	2ft6in	4wBEF 2 x15	6T Type GB6		105

HAWTHORNS & CO LTD, Leith, Edinburgh   H(L)
Steam locomotive
366   1866   4ft8½in   0-4-0WT   OC   74

R & W HAWTHORN & CO LTD, FORTH BANKS WORKS,
Newcastle upon Tyne   RWH
Steam locomotives
1665   1.1876   4ft8½in   0-6-0ST   IC   15 x 22   4ft0in 'NER Class 1350'   67
2022   .1885   4ft8½in   0-4-0ST   OC   12 x 18   3ft0in   68

R & W HAWTHORN, LESLIE & CO LTD, FORTH BANKS WORKS,
Newcastle upon Tyne   HL
Steam locomotives
2295   11.5.1895   4ft8½in   0-4-0ST   OC   14 x 20   3ft6in   68
2345   6.1896   4ft8½in   0-4-0ST   OC   14 x 20   3ft0in   68
2502   11.6.1901   4ft8½in   0-4-0ST   OC   14 x 20   3ft6in   73
2507   27.9.1901   4ft8½in   0-4-0ST   OC   14 x 20   3ft6in   78
2780   18.6.1909   4ft8½in   0-4-0ST   OC   14 x 22   3ft6in   127
2837   30.10.1910   4ft8½in   0-4-0ST   OC   14 x 22   3ft6in   73
2878   29.7.1911   4ft8½in   0-6-2T   OC   14 x 22   3ft6in   79
3460   31.3.1921   4ft8½in   0-6-0ST   OC   16 x 24   3ft8in   66,93,100
3539   9.3.1923   4ft8½in   0-4-0ST   OC   14 x 22   3ft6in   68
3540   9.6.1923   4ft8½in   0-4-0ST   OC   14 x 22   3ft6in   68
3581   23.5.1924   4ft8½in   0-4-0ST   OC   14 x 22   3ft6in   73
3632   7.4.1925   4ft8½in   0-4-0ST   OC   14 x 22   3ft6in   68
3642   25.10.1925   4ft8½in   0-6-0ST   OC   15 x 22   3ft4½in   61,88
3774   12.2.1931   4ft8½in   0-4-0ST   OC   14 x 22   3ft6in   73

ROBERT HEATH & SONS LTD, Biddulph, Stoke-on-Trent   Heath
Steam locomotive
        .1888   4ft8½in   0-4-0ST   OC   45

F. C. HIBBERD & CO LTD, Park Royal, London   FH
Petrol and Diesel locomotives

Works No	Date ex Works	Gauge	Wheel Arrgt	Horse Power	Wheel Dia	Type/Engine Make		See Page
1612 (a)	3.1929	4ft8½in	4wPM	40		Dorman		42,83,127
1846	2.1934	4ft8½in	4wPM	40	3ft1in	Dorman 4JOR	8T	42,83
1869	8.1934	2ft0in	4wDM	20	1ft5¾	National 2D	3T	47
2914	9.1944	4ft8½in	4wDM				6T *	127
3809	22.11.1963	4ft8½in	4wDM		1ft8in	Perkins L4	6½T #	76,128
3837	14.2.1958	4ft8½in	4wDM		3ft1½	Dorman 4DL	18T %	99

South Staffordshire Handbook. Page 154

* "Simplex" Type
# Type 4LS Special
% Type SCW

(a) Built by Stableford & Co Ltd, Coalville, Leicestershire.

HUDSWELL, CLARKE & RODGERS, RAILWAY FOUNDRY, Leeds   HCR

Steam locomotives

75*	30.6.1866	4ft8½in	0-4-0ST	OC	14 x 20	3ft6in		39
139	17 9.1874	4ft8½in	0-4-0ST	OC	10 x 16	2ft9in	11T	56
148	22.4.1874	4ft8½in	0-4-0ST	OC	14 x 20	3ft6in	17T	39
161	15.9.1875	4ft8½in	0-4-0ST	OC	13 x 20	3ft6in	17T	45
168	11.10.1875	4ft8½in	0-4-0ST	OC	14 x 20	3ft6in	20T	39
177	29.4.1876	4ft8½in	0-4-0ST	OC	14 x 20	3ft6in	20T	39
178	29.5.1876	4ft8½in	0-4-0ST	OC	14 x 20	3ft6in	20T	39,68
194	7.5.1878	4ft8½in	0-4-0ST	OC	13 x 20	3ft6in	18T	45

* Built by Hudswell & Clarke

HUDSWELL, CLARKE & CO LTD, RAILWAY FOUNDRY, Leeds   HC

Steam locomotives

262	27.9.1883	4ft8½in	0-4-0ST	OC	14 x 20	3ft6½in		82
272	4.1.1885	4ft8½in	0-4-0ST	OC	14 x 20	3ft6½in	19T	77,83
276	28.4.1885	4ft8½in	0-4-0ST	OC	12 x 18	3ft0in	16½T	54
319	16.5.1889	4ft8½in	0-6-0T	IC	15 x 20	3ft6½in	24T	79,82,95,112,114,115
327	16.10.1889	4ft8½in	0-6-0ST	IC	13 x 20	3ft3in	18T	122
333	20.11.1890	4ft8½in	0-6-0ST	IC	12 x 18	3ft0in		79
352	7.1.1891	4ft8½in	0-6-0T	IC	14 x 20	3ft3in		67,101
353	2.2.1893	4ft8½in	0-6-0ST	IC	14 x 20	3ft3in	22T	67
431	4.4.1895	4ft8½in	0-6-0ST	OC	14 x 20	3ft6in	22T 17c	127
452	16.1.1896	4ft8½in	0-4-0ST	OC	14 x 20	3ft6½in	22T 12c	42,82
568	10.9.1900	4ft8½in	0-6-0ST	IC	14 x 20	3ft3in	22T 12c	67
576	14.12.1900	4ft8½in	0-4-0ST	OC	14 x 20	3ft7in	20T 19c	77
602	27.12.1901	4ft8½in	0-4-0ST	OC	14 x 20	3ft6½in	22T 0c	42,82
647	3.4.1903	4ft8½in	0-4-0ST	OC	15 x 22	3ft7in	24T14c	39
690	29.1.1904	4ft8½in	0-4-0ST	OC	14 x 20	3ft6½in	21T 10c	42,82
724	26.4.1905	4ft8½in	0-4-0ST	OC	14 x 20	3ft3½in	22T 0c	77
1073	22.6.1914	4ft8½in	0-6-0ST	IC	16 x 22	3ft10in	30T 0c63,90,98,100	
1417	19.11.1920	4ft8½in	0-4-0ST	OC	14 x 20	3ft6½in	22T 10c*	42,83
1437	31.1.1921	4ft8½in	0-4-0ST	OC	15 x 22	3ft4in	24T 4c*	54
1752	10.11.1943	4ft8½in	0-6-0ST	IC	18 x 26	4ft3in	38T 0c	101,107
1822	29.4.1949	4ft8½in	0-6-0T	OC	16 x 24	3ft9in	32T 0c	127

* Weight empty

HUNSLET ENGINE CO LTD, Leeds   HE

Steam locomotives

22	12.11.1867	4ft8½in	0-4-0ST	OC	12 x 18	3ft1in	60
282	27.6.1882	4ft8½in	0-4-0ST	OC	12 x 18	3ft1in	60
397	21.4.1886	4ft8½in	0-6-0ST	IC	15 x 20	3ft4in	79
467	21.8.1888	4ft8½in	0-4-0ST	OC	15 x 20	3ft7in	60
761	26.2.1902	4ft8½in	0-6-0ST	IC	12 x 18	3ft2½in	79
1451	15.5.1924	4ft8½in	0-6-0ST	IC	16 x 22	3ft9in	67,90,101,107
1685	13.7.1931	4ft8½in	0-6-0ST	OC	14 x 20	3ft9in	90,93,95,98,101
							103,108,114,115
1800	15.6.1936	4ft8½in	0-6-0ST	IC	16 x 22	3ft9in	67,101,108
1821	30.7.1936	4ft8½in	0-6-0ST	IC	16 x 22	3ft9in	67,101,108
3772	28.4.1952	4ft8½in	0-6-0ST	IC	18 x 26	4ft3in	90,93,101,108
3776	30.9.1952	4ft8½in	0-6-0ST	IC	18 x 26	4ft3in	88,96,101,115
3777	9.10.1952	4ft8½in	0-6-0ST	IC	18 x 26	4ft3in	88
3789	25.8.1953	4ft8½in	0-6-0ST	IC	18 x 26	4ft3in	90,93,95,96
3806	22.12.1953	4ft8½in	0-6-0ST	IC	18 x 26	4ft3in	93,95,115
3807	30.12.1953	4ft8½in	0-6-0ST	IC	18 x 26	4ft3in	90.95,96
3839	30.1.1956	4ft8½in	0-6-0ST	IC	18 x 26	4ft3in	90,95,96

Diesel locomotives

2065	29.4.1940	4ft8½in	0-4-0DM	153	3ft4in	Gardner 6L3	73
2066	7.5.1940	4ft8½in	0-4-0DM	153	3ft4in	Gardner 6L3	73
2068	8.7.1940	4ft8½in	0-4-0DM	153	3ft4in	Gardner 6L3	73
2176	17.8.1940	2ft0in	4wDM	25	3ft0in	McLaren LMR2	47
4080	23.6.1950	2ft6in	0-4-0DMF	70	2ft0in	Meadows Mk 1/1	110,113
4081	30.6.1950	2ft6in	0-4-0DMF	70	2ft0in	Meadows Mk 1/1	110,113
4082	16.7.1950	2ft6in	0-4-0DMF	70	2ft0in	Meadows Mk 1/1	110,116
4083	5.10.1950	2ft6in	0-4-0DMF	70	2ft0in	Meadows Mk 1/1	110,113
4084	27.11.1950	2ft6in	0-4-0DMF	70	2ft0in	Meadows Mk 1/1	110,116
4085	23.3.1951	2ft6in	0-4-0DMF	70	2ft0in	Meadows Mk 1/1	110
4086	31.5.1951	2ft6in	0-4-0DMF	70	2ft0in	Meadows Mk 1/1	110
4087	24.3.1952	2ft6in	0-4-0DMF	70	2ft0in	Meadows Mk 1/1	1113,116
4088	18.6.1952	2ft6in	0-4-0DMF	70	2ft0in	Meadows Mk 1/1	116
4497	26.3.1956	2ft6in	0-4-0DMF	70	2ft0in		113,116
7015	5.1971	4ft8½in	0-6-0DH	400	3ft9in	Rolls-Royce C8TFL	103,108
7017	9.1971	4ft8½in	0-6-0DH	400	3ft9in	Rolls-Royce C8TFL	115
7018	10.1971	4ft8½in	0-6-0DH	400	3ft9in	Rolls-Royce C8TFL	108
7181	21.3.1970	4ft8½in	0-6-0DH	325	3ft9in	Rolls-Royce C8SFL	103,108
8825	26.10.1979	2ft6in	4wDH	52	1ft6in	Perkins 4.203	104,116
8826	21.12.1978	2ft6in	4wDH	52	1ft6in	Perkins 4.203	104
8971	11.1979	2ft6in	4wDH	52	1ft6in	Perkins 4.203	109
8973	25.01.1979	2ft6in	4wDH	52	1ft6in	Perkins 4.203	104
9041	9.8.1982	2ft6in	4wDH	52	1ft6in	Perkins 4.203	109

HENRY HUGHES & CO , Loughborough   HH

Steam locomotive

c/1875	2ft7in	0-4-0T	OC	6 x	69

JAMES C. KAY & CO LTD, PHOENIX FOUNDRY, Bury   Kay

Petrol locomotive
                 4ft8½in   4wPM   57

KENT CONSTRUCTION & ENGINEERING CO LTD, Ashford, Kent   KC

Petrol locomotives

.1924	4ft8½in	4wPM	40	3ft1in	Dorman	42,83
.1924	4ft8½in	4wPM	40	3ft1in	Dorman	42,83
.1925	4ft8½in	4wPM	40	3ft1in	Dorman	42,83
.1926	4ft8½in	4wPM	40	3ft1in	Dorman	42,83,126,127

KERR STUART & CO LTD, CALIFORNIA WORKS, Stoke-on-Trent   KS

Steam locomotive

4226	19.5.1930	4ft8½in	0-4-0ST	OC	15 x 20	3ft6in	Moss Bay	60

Diesel locomotives

4421	2.12.1929	4ft8½in	6wDM	90	18½Ton	McLaren-Benz 6cyl	76
4428	6.5.1929	4ft8½in	0-4-0DM	90	21½Ton	McLaren-Benz 6cyl	60

KITSON & CO LTD, AIREDALE FOUNDRY, Leeds   K

Steam locomotives

5036	.1913	4ft8½in	0-6-0ST	IC	16 x 22	3ft7in	49,90,93
5158	.1915	4ft8½in	0-6-0ST	IC	16 x 22	3ft7in	90
5358	.1921	4ft8½in	0-6-0T	IC	18 x 26	4ft6in	56,90,98

LILLESHALL CO LTD, Oakengates, Shropshire   Lill

Steam locomotives

.1864	4ft8½in	0-4-0ST	OC			63
.1867	4ft8½in	0-6-0ST	IC			63
.1867	4ft8½in	0-6-0ST	IC	17 x 22	3ft6in	52,90,95,100
.1868	4ft8½in	0-6-0ST	IC	17 x 22	3ft6in	52,95,115
.1868	4ft8½in	0-4-0ST	OC			52
.1870	4ft8½in	0-6-0ST	IC	17 x 22	3ft6in	52
.1872	4ft8½in	0-6-0ST	IC	17 x 21	3ft7in	52,95

R.A.LISTER & CO LTD, Dursley, Gloucestershire   L

Petrol and Diesel locomotives

962	c.1930	1ft11½in	4wPM	46
26060	.1944	1ft11½in	4wPM	46

LONDON, BRIGHTON & SOUTH COAST RAILWAY, BRIGHTON WORKS Bton

Steam locomotives
```
         3.1877    4ft8½in    0-6-0T    IC  17 x 24   4ft6in            52,95,96,126,127
```

LONDON & NORTH WESTERN RAILWAY, CREWE WORKS Crewe

Steam locomotive
```
1473     4.1872    4ft8½in    0-4-0ST   IC  14 x 20   4ft0in   24¾T              82
2979     9.1888    4ft8½in    0-6-2T    IC  17 x 24   4ft5½in  43¾T             126
```

MANNING WARDLE & CO LTD, BOYNE ENGINE WORKS, Leeds MW

Steam locomotives
```
60       28.3.1862    4ft8½in    0-4-0ST   OC   9 x 14    2ft8½in   E                  39
166      19.6.1865    4ft8½in    0-6-0ST   IC  12 x 17    3ft1³⁄₈in K         52,82,120
167      28.11.1865   4ft8½in    0-6-0T    IC  15 x 22    4ft2in    West Yorkshire  75
228      30.9.1867    4ft8½in    0-4-0ST   OC  12 x 18    3ft0in    H                  46
244      26.3.1868    4ft8½in    0-6-0ST   IC  12 x 17    3ft0in    K       54,70,90,98
458      27.7.1874    4ft8½in    0-6-0ST   IC  13 x 18    3ft0in    M                 122
556      14.6.1875    4ft8½in    0-6-0ST   IC  12 x 17    3ft1³⁄₈in K                  75
563      25.1.1876    4ft8½in    0-6-0ST   IC  12 x 17    3ft1³⁄₈in K                  67
565      12.08.1875   4ft8½in    0-6-0ST   IC  15 x 22    3ft9in    O                  54
577       8.11.1875   4ft8½in    0-4-0ST   OC   8 x 14    2ft8in    D                 120
593      20.12.1877   4ft8½in    0-4-0ST   OC  12 x 18    3ft0in    H                  46
774       7.12.1881   4ft8½in    0-6-0ST   IC  13 x 18    3ft0in    M                 122
812      18.10.1881   4ft8½in    0-6-0ST   IC  12 x 17    3ft1³⁄₈in K                  79
865      19.1.1883    4ft8½in    0-6-0ST   IC  12 x 17    3ft1³⁄₈in K                 122
973      26.5.1886    4ft8½in    0-6-0ST   IC  12 x 17    3ft1³⁄₈in K                 122
1040     25.1.1888    4ft8½in    0-4-0ST   OC  10 x 16    2ft9in    F                 121
1070     26.10.1888   4ft8½in    0-6-0ST   IC  12 x 17    3ft0in    K                 120
1145     27.06.1890   4ft8½in    0-6-0ST   IC  12 x 17    3ft0in    K                 123
1180     26.11.1890   4ft8½in    0-6-0ST   IC  15 x 22    3ft9in    O       54,70,71,90,
                                                                               95,100,107,114,115
1246      2.2.1892    2ft0in     0-4-0ST   OC   7½ x 12   2ft2in    -                  62
1326     25.3.1896    4ft8½in    0-6-0ST   IC  15 x 22    3ft9in    O   54,70,90,98,101
1371      5.6.1897    2ft0in     0-4-0ST   OC   6½ x 9    1ft8in    Special        62,75
1427      8.2.1899    4ft8½in    0-4-0ST   OC  14 x 18    3ft0in    P              46,82
1502     29.6.1900    4ft8½in    0-6-0ST   IC  12 x 17    2ft6in    K                 123
1513      2.9.1901    4ft8½in    0-6-0ST   IC  12 x 18    3ft0in    L                  79
1515     11.11.1901   4ft8½in    0-6-0ST   IC  16 x 22    4ft0in    Special    71,90,107
1596      4.3.1903    4ft8½in    0-6-0ST   IC  16 x 22    4ft0in    T             71,107
1601      8.5.1903    4ft8½in    0-6-0ST   IC  12 x 18    3ft0in    L                 123
1722     28.2.1908    4ft8½in    0-6-0ST   IC  12 x 18    3ft0in    L           70,90,98
1759     26.8.1910    4ft8½in    0-6-0ST   IC  18 x 24    4ft0in    Special       71,107
1852     24.7.1914    4ft8½in    0-6-0ST   OC  15 x 22    3ft6in    Special           45
1913      9.3.1917    4ft8½in    0-6-0ST   OC  16 x 24    3ft6in    Spl 52,71,90,95,107
2018     13.11.1922   4ft8½in    0-6-0ST   IC  18 x 24    3ft0½in   -             71,107
```

MAUDSLAY, SONS & FIELD, Lambeth Marsh, London

Steam locomotive
.1838   4ft8½in   0-4-0   IC   13 x 18   5ft0in   120

MOTOR RAIL LTD, SIMPLEX WORKS, Bedford   MR

Petrol and Diesel locomotives

Works No	Date ex Works	Gauge	Wheel Arrgt	Horse Power	Weight (Tons)	Engine Make	See Page
529	14.1.1918	2ft0in	4wPM	40	6.75		39
1947	7.4.1920	4ft8½in	4wPM	40	8		127
4519	5.6.1928	2ft0in	4wPM	20	2½		130
5206	21.8.1930	60cm	4wPM 20/35		2½		44
5324	10.6.1931	2ft0in	4wPM 20/35		2½		39
5401	13.10.1934	2ft0in	4wPM 20/35		2½	(*)	44
5630	12.9.1932	2ft0in	4wDM 20/35		2½		39
5819	7.6.1935	2ft0in	4wDM 12/16		2½		76
5825	8.6.1936	2ft0in	4wDM 16/24		2½		76
5828	27.10.1936	2ft0in	4wDM 16/24		2½		76
5829	20.11.1936	2ft0in	4wDM 16/24		2½		76
5853	18.1.1934	2ft0in	4wDM 20/28		3½		39
7170	2.4.1937	2ft0in	4wDM 20/28		2½		65
7939	1.6.1939	2ft0in	4wDM 32/42		5		72
8592	.1940	2ft0in	4wDM 20/28		2½		64,65
8681	c10.1941	2ft0in	4wDM 20/28		2½		64,65
8882	21.4.1944	2ft0in	4wDM 20/28		2½		65
21282	10.12.1959	2ft0in	4wDM 20/28		2½		64
40SD501	30.1.1975	2ft0in	4wDM	40	2½		128

*   built c/1932, converted from 4ton loco ex PLH.

LOCOTRACTEURS GASTON MOYSE, La Courneuve, Seine, France   Moyse

Petrol locomotive
(86?)   .1926   4ft8½in   4wPE   Type 20TDE   83

NEILSON & CO, Hyde Park Works, Springburn, Glasgow.   N

Steam locomotive
2937   .1882   4ft8½in   0-4-0ST   OC   14 x 20   3ft8in   127

NEILSON, REID & CO, Hyde Park Works, Springburn, Glasgow NR

Steam locomotives

5567	.1899	4ft8½in	0-4-0ST	OC	14 x 21	42
5568	.1899	4ft8½in	0-4-0ST	OC	14 x 21	42
5759	.1900	4ft8½in	0-4-0ST	OC	14 x 21	42
5760	.1900	4ft8½in	0-4-0ST	OC	14 x 21	42
5907	.1901	4ft8½in	0-4-0ST	OC	14 x 21	42,126

NORTH BRITISH LOCOMOTIVE CO LTD, Glasgow NB

HYDE PARK WORKS (NBH)
Steam locomotive

19848	.1913	4ft8½in	0-4-0ST	OC	14 x 21	3ft6in	42

QUEENS PARK WORKS (NBQ)
Diesel locomotives

27814	.1958	4ft8½in	0-4-0DH	225	3ft		76
27940	.1959	4ft8½in	0-4-0DH	225	3ft6in	NBL MAN	76

PECKETT & SONS LTD, ATLAS LOCOMOTIVE WORKS, Bristol P

Steam locomotives

440	3.11.1885	4ft8½in	0-4-0ST	OC	10 x 14	2ft6in	M3	72
567	9.10.1894	4ft8½in	0-6-0ST	IC	16 x 22	3ft10in		x56,90,101,108,115
597	102.1895	4ft8½in	0-4-0ST	OC	12 x 18	3ft0in	R1	72
618	26.3.1895	4ft8½in	0-6-0ST	IC	16 x 22	3ft10in	x	63,90,100
786	16.5.1899	4ft8½in	0-6-0ST	IC	18 x 24	4ft0½in	Q	52,95
809	7.5.1900	4ft8½in	0-6-0ST	OC	14 x 20	3ft7in	B1	66,90,100
832	21.5.1900	4ft8½in	0-4-0ST	OC	14 x 20	3ft2in	W4	60
879	21.8.1901	4ft8½in	0-6-0ST	OC	14 x 20	3ft7in	B1	82,111,114,115
917	16.1.1902	4ft8½in	0-4-0ST	OC	12 x 18	3ft0in	R1	127
937	2.4.1902	4ft8½in	0-4-0ST	OC	12 x 18	3ft0in	R1	78
1038	1.1.1906	4ft8½in	0-4-0ST	OC	12 x 18	3ft0in	R1	127
1351	8.8.1914	4ft8½in	0-4-0ST	OC	15 x 21	3ft7in	E	127
1491	5.11.1917	4ft8½in	0-4-0ST	OC	14 x 20	3ft2½in	W5	45,89,95
1585	23.10.1922	4ft8½in	0-4-0ST	OC	12 x 18	3ft0in	R2	78
1666	27.11.1924	4ft8½in	0-4-0ST	OC	12 x 18	3ft0in	R2	78
1823	28.6.1933	4ft8½in	0-4-0ST	OC	10 x 15	2ft9in	M5	127
2017	3.11.1941	4ft8½in	0-4-0ST	OC	14 x 22	3ft2½in	W7	90
2019	15.12.1941	4ft8½in	0-4-0ST	OC	14 x 22	3ft2½in	W7	80
2112	11.4.1949	4ft8½in	0-4-0ST	OC	12 x 20	3ft0½in	R4	78
2136	29.6.1953	4ft8½in	0-4-0ST	OC	12 x 20	3ft0½in	R4	78

RUSTON & HORNSBY LTD, Lincoln   RH

Diesel locomotives

166022	12.10.1933	2ft0in	4wDM	16		Lister	16HP	44
168833	30.10.1933	2ft0in	4wDM	16		Lister	16HP	64
170197	22.6.1934	2ft0in	4wDM	16		Lister	16HP	64
174548	11.4.1935	2ft0in	4wDM	20		Lister	18/21HP	44
175118	18.4.1935	2ft0in	4wDM	20		Lister	18/21HP	64
186303	17.6.1937	600mm	4wDM	40		Ruston	33/40HP	39
187056	11.11.1937	2ft0in	4wDM	13		Ruston	11/13HP	64
195868	19.6.1939	2ft0½in	4wDM	13		Ruston	11/13HP	64
218045	23.12.1942	4ft8½in	4wDM	48	2ft6in	Ruston	48DS	71
223737	14.3.1944	600mm	4wDM	20	1ft4in	Ruston	20DL	65
223747	4.1944	600mm	4wDM	20	1ft4in	Ruston	20DL	104
264242	18.8.1949	2ft0in	4wDM	13		Ruston	13DL	64
313392	2.7.1952	4ft8½in	0-4-0DM	165	3ft2½in	Ruston	165DS	121
321730	8.4.1952	4ft8½in	4wDM	88	3ft0in	Ruston	88DS	88,112
338413	28.4.1953	4ft8½in	4wDM	88	3ft0in	Ruston	88DS	88,90, 99,101,112
339267	9.7.1953	2ft6in	0-4-0DMF	75	1ft7in	Ruston	LHG	113
339268	24.7.1953	2ft6in	0-4-0DMF	75	1ft7in	Ruston	LHG	113
339275	30.10.1953	2ft6in	0-4-0DMF	75	1ft7in	Ruston	LHG	113
370546	28.5.1954	2ft6in	0-4-0DMF	75	1ft7in	Ruston	LHG	102
370552	16.7.1954	2ft6in	0-4-0DMF	75	1ft7in	Ruston	LHG	102
374449	17.9.1954	2ft6in	0-4-0DMF	75	1ft7in	Ruston	LHG	102
374452	13.10.1954	2ft6in	0-4-0DMF	75	1ft7in	Ruston	LHG	113
386873	25.4.1955	4ft8½in	4wDM	48	2ft6in	Ruston	48DS	71
388770	1.6.1955	2ft0in	0-4-0DMF	75	1ft7in	Ruston	LHG	89
388773	27.7.1955	2ft6in	0-4-0DMF	75	1ft7in	Ruston	LHG	102
392155	15.6.1956	2ft6in	0-4-0DMF	75	1ft7in	Ruston	LHG	113
393303	20.1.1956	4ft8½in	4wDM	48	2ft6in	Ruston	48DS	79
402815	18.9.1956	2ft0in	4wDM	48		Ruston	48DL	130
412716	1.11.1957	4ft8½in	0-4-0DE	210	3ft2½in	Ruston	200DE	42
416566	6.11.1957	4ft8½in	4wDM	88	3ft0in	Ruston	88DS	42
432664	24.8.1959	2ft0in	4wDM	31½	1ft4in	Ruston	LBU	64
441945	24.12.1959	2ft6in	4wDM	31½	1ft4in	Ruston	LBT	116
441946	24.12.1959	2ft6in	4wDM	31½	1ft4in	Ruston	LBT	109,116
441947	30.12.1959	2ft6in	4wDM	31½	1ft4in	Ruston	LBT	92,104,118
441948	30.12.1959	2ft6in	4wDM	31½	1ft4in	Ruston	LBT	104,112
452293	25.11.1960	2ft6in	4wDM	31½	1ft4in	Ruston	LBT	96,109
458641	31.1.1963	4ft8½in	0-4-0DE	165	3ft2½in	Ruston	165DE	127
466587	7.11.1961	2ft6in	4wDM	31½	1ft4in	Ruston	LBT	92,96,104,109
476107	20.7.1964	2ft6in	4wDM	48		Ruston	LFT	109
476112	27.4.1962	2ft6in	4wDM	31½	1ft4in	Ruston	LBU	102,104
497760	16.9.1963	2ft6in	4wDM	44		Ruston	48DLG	116
506491	26.3.1964	2ft6in	4wDM	31½	1ft4in	Ruston	LBT	104,109
7002/0767/6	3.2.1967	2ft6in	4wDM	31½	1ft4in	Ruston	LBT	104,116
7002/0867/3	10.2.1967	2ft6in	4wDM	31½	1ft4in	Ruston	LBT	90,102,116

ROLLS-ROYCE LTD, SENTINEL WORKS, Shrewsbury   RR

Diesel locomotives
10240	09.12.1966	4ft8½in	0-6-0DH	325	3ft6in	Rolls Royce C8SFL 48T	103
10255	25.03.1966	4ft8½in	0-6-0DH	325	3ft6in	Rolls Royce C8SFL 48T	103

ROBERT STEPHENSON & CO LTD, FORTH BANKS WORKS
Newcastle upon Tyne   RS

Steam locomotives
630	12.1848	4ft8½in	0-6-0ST	IC	18 x 24	5ft0in	63
631	12.1848	4ft8½in	0-6-0ST	IC	18 x 24	5ft0in	63

ROBERT STEPHENSON & HAWTHORNS LTD, FORTH BANKS WORKS,
Newcastle upon Tyne   RSHN

Steam locomotives
7106	20.10.1943	4ft8½in	0-6-0ST	IC	18 x 26	4ft3in	49,90,93,108
7292	19.07.1945	4ft8½in	0-6-0ST	IC	18 x 26	4ft3in	90,107

Diesel and Battery locomotives
7945	.1959	2ft6in	4wBEF	64	2ft0in	EM2B1	110
8129	.1959	2ft6in	4wBEF	64	2ft0in	EM2B1	110
8131	.1959	2ft6in	4wBEF	64	2ft0in	EM2B1	97,110
8132	.1959	2ft6in	4wBEF	64	2ft0in	EM2B1	97,117
8133	.1959	2ft6in	4wBEF	64	2ft0in	EM2B1	97,105
8134	.1960	2ft6in	4wBEF	64	2ft0in	EM2B1	97,117

ROBERT STEPHENSON & HAWTHORNS LTD,
DARLINGTON WORKS   RSHD

Battery locomotives
8203	1960	2ft6in	4wBEF	64	2ft0in	EM2B1	97,110
8204	1960	2ft6in	4wBEF	64	2ft0in	EM2B1	105
8205	1960	2ft6in	4wBEF	64	2ft0in	EM2B1	110
8206	1960	2ft6in	4wBEF	64	2ft0in	EM2B1	110
8288	1961	2ft6in	4wBEF	64	2ft0in	EM2B1	105
8289	1961	2ft6in	4wBEF	64	2ft0in	EM2B1	105
8291	1961	2ft6in	4wBEF	64	2ft0in	EM2B1	106,110
8292	4.4.1961	2ft6in	4wBEF	64	2ft0in	EM2B1	110
8296	1961	2ft6in	4wBEF	64	2ft0in	EM2B1	116
8297	1961	2ft6in	4wBEF	64	2ft0in	EM2B1	110,116
8302	1962	2ft6in	4wBEF	64	2ft0in	EM2B1	116
8344	1962	2ft6in	4wBEF	64	2ft0in	EM2B1	105
8345	1962	2ft6in	4wBEF	64	2ft0in	EM2B1	105
8420	1963	2ft6in	4wBEF	64	2ft0in	EM2B1	105
8421	1963	2ft6in	4wBEF	64	2ft0in	EM2B1	106,116

SENTINEL (SHREWSBURY) LTD, Shrewsbury S

Steam locomotives
6661	.1926	4ft8½in	0-4-0VBT VCG	6.75 x 9 2ft6in	100		61
9376	.1947	4ft8½in	4wVBT VCG	6.75 x 9 2ft6in	100		68
9384	.1947	4ft8½in	4wVBT VCG	6.75 x 9 2ft6in	100		68
9632	15.4.1957	4ft8½in	4wVBT VCG	6.75 x 9 2ft6in	100		127

Diesel locomotives
10003	7.5.1959	4ft8½in	4wDH	230	3ft2in	Rolls Royce C6SFL	42
10085	28.9.1961	4ft8½in	4wDH	230	3ft2in	Rolls Royce C6SFL	42

SHARP STEWART & CO LTD, ATLAS WORKS, Manchester SS

Steam locomotive
2643	1876	4ft8½in	0-6-0ST	IC	16 x 24	49,93

JOHN SMITH, VILLAGE FOUNDRY, Coven, near Wolverhampton. JS

Steam locomotive
	c.1869	4ft8½in	0-6-0T	13 x	63

THORNEWILL & WARHAM, Burton-on-Trent TW

Steam locomotives
223	.1863	4ft8½in	0-4-0WT	OC			39
224	.1863	4ft8½in	0-4-0WT	OC			41
249	.1864	4ft8½in	0-4-0WT	OC			41
259	.1864	4ft8½in	0-4-0WT	OC			41
	.1867	4ft8½in	0-4-0WT	OC			68
303	.1869	4ft8½in	0-4-0WT	OC			41
	.1869	4ft8½in	0-4-0WT	OC	12 x 16		77
	.1872	4ft8½in	0-4-0ST	OC			45
353	.1872	4ft8½in	0-4-0WT	OC			41
373	.1873	4ft8½in	0-4-0WT	OC	14 x 20	4ft0in	41
393	.1874	4ft8½in	0-4-0WT	OC	14 x 20	4ft0in	39,41
400	.1875	4ft8½in	0-4-0WT	OC	14 x 20	4ft0in	41
420	.1876	4ft8½in	0-4-0WT	OC	14 x 20	4ft0in	41
425	.1877	4ft8½in	0-4-0WT	OC	14 x 20	4ft0in	41
455	.1880	4ft8½in	0-4-0ST	OC	14 x 20	4ft0in	42
609	.1890	4ft8½in	0-4-0ST	OC	14 x 20	4ft0in	42

THE VULCAN FOUNDRY LTD, Newton le Willows, Lancs VF

Diesel locomotive
4861	.1942	4ft8½in	0-4-0DM	153	3ft3in	Gardner 6L3	73

D WICKHAM & CO LTD, Ware, Herts   Wkm
Diesel traction
7346   29.10.1957   4ft8½in   4w-4wDMR300   2 x Leyland RC680   127

YORKSHIRE ENGINE CO LTD, MEADOW HALL WORKS, Sheffield   YE
Steam locomotives
185   1872   4ft8½in   2-4-0T   OC   15 x 22   5ft0in   52,95
406   1886   4ft8½in   0-4-0ST   OC   12 x 20   3ft3in   Class Dx   78
1027   1910   4ft8½in   0-4-0ST   OC   14 x 20   3ft3in   60
Diesel locomotives
2748   11.09.1959   4ft8½in   0-6-0DE   400   Janus   2 x R-R C6SFL   101,108
2749   14.09.1959   4ft8½in   0-6-0DE   400   Janus   2 x R-R C6SFL   101,108

INDEX OF PROPRIETORS

A

Albion Brewery (Burton on Trent) Ltd	134
Alders (Tamworth) Ltd.	38
Samuel Allsopp & Sons Ltd.	38
Amington Colliery Co	60
Marquis of Anglesey	47,134

B

Baggeridge Brick & Tile Co Ltd.	39
E E Baguley	130
Baguley Cars Ltd.	130
Baguley Engineers Ltd.	130
Baguley Drewry Co Ltd.	130
Bass & Co	40
Bass Ltd.	126
Bass, Mitchells & Butlers Ltd.	40
Bass, Ratcliff & Gretton Ltd.	40
W H Beaumont	139
Edward Best	69
Birmingham Banking Co	139
Birmingham & Liverpool Junction Canal	134
Birmingham Canal Navigation	43,45
Birmingham Railway Carriage & Wagon Co Ltd.	79
H Blewitt	140
S Blewitt & Co Ltd	140
Braddock and Matthews	120
A Braithwaite and Co	120
Branston Artificial Silk Co Ltd.	44
Branston Gravels Ltd.	44
Thomas Brassey	120
Brereton Collieries Ltd.	44
William Briers and Sons (Tamworth)	130
British Coal Corporation	85
British Electricity Authority	135
British Waterways	45
Burton Brewery Co Ltd.	46
Burton Constructional Engineering Co Ltd.	46
Burton Corporation.	135,136

C

Cannock Chase Colliery Co Ltd.	47
Cannock Chase Railway	50
Cannock Chase & Wolverhampton Railway	50
Cannock & Huntington Colliery Co Ltd.	135
Cannock & Leacroft Colliery Co	135
Cannock Lodge Colliery Co Ltd.	71
Cannock & Rugeley Colliery Co Ltd.	51

Cannock & Wimblebury Colliery Co Ltd.	53
Charrington Co Ltd.	53
Central Electricity Authority	135
Central Electricity Generating Board	135
Chasewater Light Railway and Museum Co Ltd.	126
Conduit Colliery Co Ltd.	54
Thomas Cooper	140
Coppice Colliery Co Ltd.	55
Croda Hydrocarbons Ltd	56
Croda Synthetic Chemicals Ltd.	56
Crosse & Blackwell Ltd.	56
N Curtis	128
Cyclops Engineering Co Ltd.	57

D

Darlaston Coal & Iron Co Ltd.	57
Darlaston Steel & Iron Co Ltd.	57
J and T Dumolo.	69

E

James Eadie	136
East Cannock Colliery Co Ltd.	136
East Midlands Gas Board.	136
Essington Farm Colliery Co.	58,136
Everard & Co	136
James Evers-Swindell	139

F

The Fair Oak Colliery Co Ltd.	59
Ferro (Great Britain) Ltd.	137
W C & H O Firmstone	60
Matthew Frost.	137

G

J J Gallagher (London) Ltd.	123
J.Gibb & Sons.	137
Gibbs & Canning Ltd.	59
Bernard Gilpin.	61,137
William Gilpin senior & Co Ltd.	138
Glascote Colliery Co Ltd.	60
Grant, Lyon Eagre Ltd.	121
GWR Engineers Department	121
Great Wyrley Colliery Co Ltd.	61
Greenwood & Co	142
Greenwood & Sinclair	142

H

R. O. Hanbury	55
H & E W Harrison	44
William Harrison Ltd.	62,138
Haunchwood Lewis Brick & Tile Co Ltd.	64
Henry Hawkins.	138
Joseph Hawkins & Sons.	65
T.A.Hawkins & Sons.	65
Hilton Main & Holly Bank Colliery Co Ltd.	66
Hockley hall Colliery Co Ltd	139
Hodges & Porter	130
Holly Bank Coal Co Ltd.	66

I

Ind Coope Ltd.	67,139
Ind Coope & Allsopp	67,139

K

Kettlebrook Colliery Co Ltd.	69

L

George Law	121
J.T.Leavesley.	131
G.W.Lewis Tileries Ltd.	64
Lichfield District Council	128
The Littleton Collieries Ltd.	70
Littleworth Extension Railway.	69
Lloyds (Burton) Ltd.	71
F.H.Lloyd & Co Ltd.	71
Sarah Lounds	71
William Lounds	71
Henry Lovatt	122
T Lowe	74

M

Mann, Crossman & Paulin	134
Marston, Thompson & Evershed Ltd.	72
Sir Alfred McAlpine.	72
McClean and Chawner	47
McKenzie, Stephenson & Brassey	139
Mid Cannock Colliery Co Ltd.	62,140
Midland Joinery Works Ltd	140
Midland Tar Distillers Ltd.	56
Midland - Yorkshire Tar Distillers Ltd.	56
E J Miller & Co Ltd	140
Samuel Mills	57
Ministry of Aviation	80
Ministry of Defence	73
Ministry of Munitions	74
Ministry of Supply	71
J.Murphy & Sons Ltd.	131

N
National Coal Board	85
New Cannock & Wimblebury Colliery Co Ltd.	74
Nook & Wyrley Colliery Co Ltd	62,75
Norton Cannock Colliery Co Ltd.	75

O
J Owen & Co	55

P
Joseph Palmer	138
Pauling & Co Ltd	122
Perrens and Harrison	75
Perry & Co (Bow) Ltd.	122
Francis Piggott	140
Thomas Piggott	43,48

Q
Quinton Colliery Co Ltd.	140

R
Railway Preservation Society	126
River Trent Catchment Board	76
Thomas Robinson & Co.	141
Robinson's Brewery Ltd	141
Rom Ltd	76
Rom River Plasclip Ltd	76
Rom River Reinforcement Ltd.	76
Royal Ordnance Factory	80
Rykneild Engine Co Ltd	130

S
Thomas Salt & Co Ltd	77
E Sayer	140
Earl of Shrewsbury's Brereton Collieries Ltd.	44
George Skey & Co Ltd.	77,141
Staffordshire & Worcestershire Canal	141
John Stanley	141
Stephenson Clarke Ltd.	99
A Streeter & Co Ltd.	123
J Strike	128

T
Lord Talbot	44
Tame Valley Colliery Co	142
Thornewill & Warham.	131
Trueman Hanbury Buxton & Co Ltd.	78

V
H.Vernon　　　　　　　　　　　　　142

W
Wagon Repairs Ltd.　　　　　　　　79
Peter Walker & Son　　　　　　　　142
War Department　　　　　　　　73,79,90
West Cannock Colliery Co Ltd.　　　81
Henry & Thomas Wilders　　　　　　46
Benjamin Wilson　　　　　　　　　38
Wolverhampton & Cannock Chase Railway　142
Woods & Greenwood　　　　　　　142
Worthington & Co Ltd.　　　　　　　82
Wyrley Cannock Colliery Co Ltd.　　　143

INDEX OF LOCATIONS

Entries in *italics* refer to non-locomotive lines

A

Abbey Brewery	53
Albion Brewery	*134*
Alders Paper Mills	38
Allsopp's Brewery	38,68
Alrewas stores	131
Amington Colliery	60,88
Anderstaffe Lane Maltings	40,77
Anglesey Sidings	47

B

Baggeridge Brick Works	39
Baggeridge Colliery	88
Bass Museum	126
Beacon Park, Lichfield	128
Belfast Pit	45
Black Eagle Brewery	78
Branston Works	44,56,74,79
Branston Gravel pits	44
Branston Ordnance Depot	73
Brereton Collieries	44,89
Brick Kiln Pit	45
Brindley Heath Colliery	81
Brownhills Collieries	55,62
Brownhills Common Tramroad	*134*
Brownhills railway	122
Brownhills Tramroad	*135*
Brownsfield Colliery	*137*
Burton Gasworks	*136*
Burton Power Station	*135*
Burton-on-Trent Brewery	40,46,67,68,72,77,82
Burton-on-Trent Works	130

C

Cathedral Pit	62
Cannock & Leacroft Colliery	94
Cannock & Rugeley Collieries	94
Cannock Central Stores	90
Cannock Central Workshops	90
Cannock Chase Collieries	47,92
Cannock Chase Military Railway	79
Cannock Chase Railway	49
Cannock Lodge Colliery	71
Cannock Mineral Railway	120
Cannock Old Coppice Colliery	65,94,100
Cannock Wood Colliery	51,53,94,98
Chasewater Light Railway	127
Cheslyn Hay Tramroad	*141*

Cheslyn Hay Works	64
Churchbridge Colliery	*137*
Churchbridge Works	*137*
Clarence Brewery	*142*
Conduit Colliery	70,98
Conduit Pits	55
Coppice Colliery	55,98
Coppice Pits	45,55
Coven contract	122
Crescent Brewery	*140*
Cross Street Brewery	*136*
Curzon Street Brewery	*139*

D

Dallow Lane Branch	120
Derby Road Works	46
Dixie Stores	40

E

East Cannock Colliery	99,*136*
Elford Workshops	76
Essington Colliery	*136,142*
Essington Disposal Point	99
Essington Wood Colliery	57,66
Essington Farm Collieries	58,*136*
Essington Works	64

F

Fair Oak Colliery	59
Featherstone Factory	80
Four Ashes Works	56

G

Glascote Colliery	60
Glascote Scrapyard	130
Glascote Tileries	59
Great Wyrley Colliery	61,75
Grove Colliery	62,100
Guild St Brewery	40

H

Hammerwich Colliery	47
Hatherton Colliery	136,139
Hawks Green Lane Plant Depot	131
Hawkins Colliery	100
The Hayes Pit	45,*134*
Hednesford Colliery	43,48,81
Hednesford Depot	126

High St Brewery	40
Hilton Colliery	*139*
Hilton Main Colliery	66,101
Hockley Hall Colliery	*139*
Holly Bank Colliery	66,101

I
Ind Coope Brewery	68

J
Jerome Colliery	54
Joinery Works	*140*

K
Kettlebrook Colliery	69
Kingswinford Branch	121,122
Kinver Light Railway	121

L
Leacroft Colliery	*135*
Lea Hall Colliery	103
Lichfield contract	123
Littleton Colliery	70,107,121,*135*
Littleworth Extension Railway	47,51,69
Littleworth Tramway	43
Longhouse Colliery & Brickworks	*138*

M
Mid Cannock Colliery	63,112
Middle Brewery	40
Milford Gravel Pits	72

N
New Brewery	38,40
New Street Works	131
New Wyrley Colliery	*137*
Nook & Wyrley Colliery	113
Norton Canes Canal Wharf	45
Norton Cannock Colliery	75
Norton Extension Railway	*137*
Norton Green Colliery	54

O
Old Brewery	38,40
Old Coppice Colliery	65,113
Old Engine Pit	45
Old Falls Colliery	143

P
Paradise Factory	80
Peel Colliery	78
Penkridge contract	120
Pennyfield Colliery	*139*
Pool Colliery	51

Q
Quinton Colliery	140

R
Rawnsley Depot	51,113
Rosemary Tileries	64
Rugeley School Colliery	*137*

S
Shobnall Brewery	*142*
Shobnall Maltings	40, *139*
Sneyd Colliery	65
The Springs Brickworks	45
Springhill Colliery	58

T
Tame Valley Colliery	77
Trent Brewery	*136*
Trent Valley Works	76

U
Union St Brewery	*141*
Uxbridge Pit	47

V
Valley Colliery	51,53,94,113
Valley Training Centre	97
Victoria Street Works	57

W
Wellington Works	71
West Cannock Collieries	76,94,114,115
Wilnecote Brickworks	78
Wilnecote Colliery	75
Wimblebury Colliery	51,53,74,94,117
Wombourn Works	*137*
Wyrley Cannock Collieries	*143*
Wyrley Collieries	63,100,118,*137*,*143*

Y
Yew Tree Drift Mine	118

AB 1576/1918 0-6-0ST OC   [R K Hateley]
Alders Paper Mills, Tamworth, 19/6/1959. One of two identical locomotives originally supplied in 4/1918 to Nobels Explosives, Pembrey, Carmarthenshire.

MW 556/1875 0-6-0ST IC CANNOCK No.1   [Jim Evans]
A general view of the yard, wagons and locomotive at Norton Cannock Colliery, near Bloxwich.

A general view of West Cannock No.1 Colliery which shows two locomotives.

JS /1869 0-6-0ST SUCCESS [R T Russell]
William Harrison Ltd. A group of colliery workers line up in front of the engine.

HC 353/1893 0-6-0ST IC HOLLYBANK No.1 [R T Russell]
Hollybank Colliery Co. No.1 was one of two Hudswell Clarke locomotives which handled the colliery traffic at the turn of the century.

TW 224/1863 0-4-0WT OC [F Jones]
This early Thornewill & Warham locomotive was the first to be used by Bass on their internal brewery railway system. Several other locomotives were built to this design.

TW 609/1890 0-4-0ST OC NO.3 [F Jones]
Bass, Ratcliffe & Gretton Ltd. No 3 was one of two 'Triangle' class locomotives which were supplied as saddle tank engines.

NR 5760/1900 0-4-0ST OC No.2 [K J Cooper]
Bass, Ratcliffe & Gretton Ltd. This was one of six locomotives (Class A) built by Neilson or their successors which were used on this system

HC 272/1885 0-4-0ST OC No.1 [F Jones]
No.1 worked at Salt's Brewery up to its closure in 1927.

HCR 168/1875 0-4-0ST OC No.4 [F Jones]
Samuel Allsopp & Sons Ltd. Allsopp was the first of the Burton brewery owners to use his own locomotives.

HC 1417/1920 0-4-0ST OC No.6   [K J Cooper]
Worthington & Co Ltd. The firms of Allsopp and Worthington preferred HC locomotives for their railways. No.6 is seen at the brewery in 1950.

KC /1924 4wDM No.7   [K J Cooper]
Worthington & Co Ltd, c/1949. No.7 was the first of a number of internal combustion locomotives to be used on Worthington's brewery traffic.

WB 2815/1945 0-4-0ST OC No.1 [K J Cooper]
Worthington & Co Ltd, 15/4/1952. No.1 is seen at The Hay.

HC 602/1901 0-4-0ST OC No.15 [R K Hateley]
From 1956 the locomotives operated by Bass and Worthington were pooled into one fleet. No.15, earlier Worthington No.4 was at Bass' Guild Street shed on 27/5/1961.

HL 3450/1923 0-4-0ST OC No.4   [K J Cooper]
Ind Coope & Allsopp Ltd, 1/7/1950. Thie firm of Ind Coope preferred HL locomotives.
No.4 came from Romford Brewery after the merger with Samuel Allsopp.

Bg 3227/1951 0-4-0DM No.2   [R K Hateley]
Ind Coope Ltd, 24/4/1964.

EE 533/1922   4wBE   No.9   [R K Hateley]
Ind Coope Ltd, 24/4/1964.   Standard gauge battery locomotives such as this were rarely found in industrial service.

HL 3774/1931   0-4-0ST   OC   No.4   [K J Cooper]
Marston, Thompson & Evershed Ltd, 15/4/1952.   No. 4 worked at Marston's brewery for its entire existence.   She was scrapped in 1955.

Bg 3410/1955 0-4-0DM No.4 [Baguley-Drewry]
Marston, Thompson & Evershed Ltd. When HL 3774 was scrapped, parts were incorporated into this new Baguley locomotive.

P 809/1900 0-6-0ST OC HAWKINS [K J Cooper]
NCB Hawkins Colliery (Old Coppice), 22/4/1958. HAWKINS, named after the mine owning family, was the first engine at this colliery and remained here until scrapped.

SS 2643/1876 0-6-0ST IC No.6 [K J Cooper]
NCB Cannock Chase Colliery, 7/4/1950. No.6 was often employed on the duties farthest from Chasetown and ran with an ancilliary four wheel wooden tender.

HE 3789/1953 0-6-0ST IC No.3 [K J Cooper]
NCB Cannock Chase Colliery 22/3/1953. This standard Hunslet 'Austerity' locomotive is seen outside the shed soon after its delivery.

RWH 1665/1876 0-6-0ST IC No.10 [F Jones]
Hilton Main & Hollybank Collieries Ltd. No.10 stands in front of a group of private owner wagons on the Hilton Main & Hollybank Colliery railway.

HE . 1821/1936 0-6-0ST IC CAROL ANN No.5 [K J Cooper]
NCB Hollybank Colliery, 31/3/1951. No.5 was one of two identical Hunslets bought by Hilton Main & Hollybank Collieries Ltd in 1936 to replace older locomotives.

MW 1596/1903 0-6-0ST IC LITTLETON No.2 [K J Cooper]
NCB Littleton Colliery, 8/2/1958. The Littleton Colliery Co preferred to buy MW locomotives and No.2 was one of two 16-inch engines built for the mine's opening.

HC 1073/1914 0-6-0ST IC THE COLONEL [K J Cooper]
NCB Grove Colliery. THE COLONEL is named after Colonel W E Harrison, a director of William Harrison Ltd, one-time owner of Grove Colliery.

BP 28/1856 0-4-2ST IC McCLEAN [K J Cooper]
NCB Cannock Chase, 7/4/1950. McCLEAN is seen in steam at the end of a very long working career. It was the first of five locomotives of this type used at this location.

Chasetown /1946 0-4-2ST IC FOGGO [K J Cooper]
NCB Cannock Chase. FOGGO was built at Chasetown from parts of old BP locomotives and new components from BP and was named after manager M Foggo.

HE 1685/1931 0-6-0ST IC NUTTALL   [K J Cooper]
NCB Cannock Wood, 2/3/1963. NUTTALL came to the NCB from the contractors Mowlem and was used at many collieries as the area spare locomotive.

HC 319/1889 0-6-0T IC STAFFORD   [K J Cooper]
NCB Hednesford Colliery, 11/10/1952. STAFFORD was also at first a contractors locomotive, coming to the Cannock Chase Military Railway and then to colliery service.

Lill /1868 0-6-0ST IC MARQUIS [K J Cooper]
NCB Rawnsley Shed. The Cannock & Rugeley Colliery Co employed a varied fleet of locomotives, including several from the Shropshire firm of Lilleshall & Co.

Lill /1868 0-6-0ST IC ANGLESEY [K J Cooper]
NCB Rawnsley Shed. The first locomotives purchased for this line remained here for over ninety years. Their names derived from the landowner, the Marquis of Anglesey.

YE 185/1872 0-6-0T OC HARRISON   [K J Cooper]
NCB Wimblebury Colliery, /1952. Built for the Potteries, Shrewsbury & North Wales Rly as a 2-4-0T, HARRISON had a varied career including work on the East & West Jct Rly.

Rawnsley, /1888 2-4-0T IC BIRCH   [K J Cooper]
NCB Brereton Colliery, 23/11/1957. The Cannock & Rugeley Colliery Co workshops at Rawnsley were sufficiently extensive to build this locomotive from their own resources.

HE 3807/1953 0-6-0ST IC 8 [K J Cooper]
NCB Rawnsley Shed, 2/3/1963. No.8 was new here in 1953 but was scrapped after a short life of only fourteen years.

MW 244/1867 0-6-0ST IC CONDUIT No.1 [K J Cooper]
This early MW is seen out of use at Chasetown Workshops, 20/6/1950, after a working life spent entirely at Conduit Colliery.

P 567/1894 0-6-0ST IC HANBURY [K J Cooper]
NCB Coppice Colliery, 22/3/1958. HANBURY was supplied new to Coppice Colliery where it remained until frequent transfers between pits in later NCB days.

GECT 5421/1977 6wDE WESTERN ENTERPRISE [R L Waywell]
NCB Littleton Colliery, 6/6/1979. Locomotives of this 750hp design now exclusively handle traffic to Penkridge from this last working mine in South Staffordshire.

RH 506491/1964 4wDM Gauge: 2ft 6in   [R K Hateley]
NCB Littleton Colliery, 17/8/1978. From 1959 the South Staffordshire collieries worked
surface stockyard railways with locomotives, such as this Ruston LBT.

RH 452293/1960 4wDM Gauge: 2ft 6in   [R M Shill]
NCB Cannock Wood Colliery. A slightly battered Ruston LBT trundles towards the
stockyard past the old colliery buildings.

HE 8973/1979 4wDH Gauge: 2ft 6in [P Fidczuk]
One of the five 52hp Hunslet locomotives supplied to South Staffordshire mines, seen here on the stockyard railway at Lea Hall Colliery.

GB 6090/1963 4wBEF Gauge: 2ft 6in [P Fidczuk]
This battery locomotive, normally used underground, is seen on surface at Lea Hall Colliery.

CE 5074/1966 (right) and CE 5097/1966 (left), 4wBEF, Gauge 2ft 6in   [P Fidczuk]
Lea Hall Colliery. These were former underground locomotives later used on the stockyard railway.

[E E Pritchard]
An unidentified Ruston LHG locomotive stands at Leacroft Junction underground at Mid-Cannock Colliery in 11/1953.

RR 10255/1966 0-6-0DH [R L Waywell]
British Coal, Lea Hall Colliery, 4/9/1990. The working life of this modern colliery, much of whose output went to the power station in the background, was short.

EEV D1120/1966 0-6-0DH [K Lane]
NCB, Lea Hall Colliery, 8/12/1977. This locomotive, seen beside the colliery screens, was first used at Cannock Wood Colliery.

MW 774/1881 0-6-0ST IC DAVID [E Milner]
Perry & Co, Kingswinford contract. DAVID is seen with contractors wagons at the Planks Lane shed and workshops at Wombourn.

KS 4421/1929 6wDM [K Lane]
Rom River Reinforcement Co Ltd, 8/12/1977. This early diesel locomotive first worked on the Ravenglass & Eskdale Rly and later in Co. Durham.